MUMFORD,
TATE,
EISELEY

MUMFORD, TATE, EISELEY

WATCHERS IN THE NIGHT

Gale H. Carrithers, Jr.

LOUISIANA STATE UNIVERSITY PRESS

BATON ROUGE AND LONDON

First printing
00 99 98 97 96 95 94 93 92 91   5 4 3 2 1

Designer: Rebecca Lloyd Lemna
Typeface: Sabon
Typesetter: G & S Typesetters, Inc.
Printer and binder: Thomson-Shore, Inc.

Library of Congress Cataloging-in-Publication Data
Carrithers, Gale H., date.
     Mumford, Tate, Eiseley : watchers in the night / Gale H.
Carrithers, Jr.
       p.   cm.
     Includes bibliographical references and index.
     ISBN 0-8071-1650-5
     1. American literature—20th century—History and criticism.
2. Literature and technology—United States—History—20th century.
3. Technology and civilization in literature.   4. Technology—Social
aspects—United States.   5. Mumford, Lewis, 1895–   —Views on
technology and civilization.   6. Eiseley, Loren C., 1907–1977—
Knowledge—Technology.   7. Tate, Allen, 1899–   —Knowledge—
Technology.   I. Title.
PS228.T42C3   1991
810.9'356'0904—dc20   91-11860
                                                              CIP

"Colonel Tate, in Attack and Defense" is reprinted, in revised form, from *Southern Literature and Literary Theory,* ed. Jefferson Humphries, copyright © 1990, University of Georgia Press; used by permission. Grateful acknowledgment is offered for permission to quote from the poems of Loren Eiseley: Excerpts from "All the Night Wings" reprinted with permission of the Estate of Loren Eiseley from *All the Night Wings,* by Loren Eiseley. Excerpts from "A Hider's World," "The Little Treasures," "Star with a Secret," and "Two Hours from Now" reprinted with permission of Charles Scribner's Sons, an imprint of Macmillan Publishing Company, from *Another Kind of Autumn,* by Loren Eiseley. Copyright © 1977 by the Estate of Loren Eiseley. "Two Hours from Now" originally appeared in *Poetry.*

To S.J.C.,

M.C.S.,

and C.H.S.

For we wrestle not against flesh and blood, but against principalities, against powers, against the rulers of the darkness of this world, against spiritual wickedness in high places.

—Paul to the Ephesians 6:12

We have enough to do to make up ourselves from present and passed time, and the whole stage of things scarce serveth for our instruction.

—Sir Thomas Browne, Dedicatory Letter to *Hydriotaphia*

Paper becomes the chief medium of embodiment: to become real was to exist on printed paper.

—Lewis Mumford, *The Condition of Man*

CONTENTS

PREFACE

For all the particular blessings that have come to us, who among us has not worried that in a general way things may be going bad? Even apart from the menace of nuclear war, may not things get uncontrollably worse in some confluence of social and environmental senses? What in general is *wrong?* What in general should we do? How should we be?

An awakening to such questions began for me when I read Lewis Mumford's urban pieces in the *New Yorker* during my high-school and early college years. Mumford's strictures on the American love affair with the automobile especially troubled me: how could he be so harsh about one of the two or three most intriguing things in the world? Loren Eiseley, with his own defamiliarizing ways of looking at the Western world in general, and at modes of perception from the animal to the scientific to the hermeneutic, was for me a later discovery. Still later, living a second time in the South, I returned to Allen Tate, whose essays and poetry I had read uncomprehendingly in graduate school. I returned to him as my curiosity about southern perspectives deepened and my suspicion grew that Mumford, Eiseley, and he triangulated a profound answer to the question of what is wrong—of, in Tate's phrase, the "deep illness of the modern mind." I also came to see that the essays of the three men manifested a way, more general than writing, of being apart from and even countering that illness.

This is not a biographical study. Others have done and are doing that important work: Donald Miller for Mumford (too recently for discussion here), Gale Christianson just now for Eiseley, others late and soon for Tate. Still, my study has a biographical side in bringing out one of the important congruities of the three men: a certain kind of American identity, marginal, ambivalently engaged in a culture they partly excoriated yet by whose fellow participants they wanted to be loved, and uneasy with institutional appointments, however convenient academic or other posts might have been as vantage points and work stations.

A second congruity of the three lies in the way they are most characterized by the profound mediation of the prophetic essay and, arguably, are most richly themselves in that literary form. In the prophetic essay, they somewhat similarly detach themselves from and at the same time connect themselves to their neighbors. Along with that, there is no bald, unproblematic version of their composite "answer" which can be thought of as preexisting their writings and no adequate version that can be leached or dissected out of their writings. I can point, register, invoke. But the doing, in each body of essays, is the saying. The writing is each essayist's witness to the "quarrel with ourselves" and his act of resistance to the "mad abstract dark"—in two Yeatsian phrases that Tate honors. I shall give less exclusively rhetorical or linguistic attention to these essayists' ways than some scholars would expect, but more than some others with a historical focus can relish. Edward Said has recognized a recent period of "almost Renaissance brilliance" on the part of "oppositional, or avant-garde literary theory," but he has complained that "nowhere in all this will one encounter a serious study of what authority is, either with reference to the way that authority is carried historically and circumstantially from the State down into a society saturated with authority of one sort or another, or with reference to the actual workings of culture, the role of intellectuals, institutions, and establishments."[1] Precisely those workings and roles concern Mumford, Tate, and Eiseley, who typically meet them at the earlier stages of the etiology of false authority or misapplied power and articulate what they find in tropes—sometimes implicit—other than Said's "from the State down."

The three prophets address the mind, as Tate insisted and the other two acknowledged a bit less directly—the mind in its formulative and reflective modes somewhat short of action and direction. They concern themselves with an agglomeration of notions diffusing in all directions, coming to be reified and reinforced in institutions, dangerous partly because so often too amorphously unconscious to be objectified as ideas

1. Edward Said, "Reflections on Recent American 'Left' Literary Criticism," in *The Question of Textuality: Strategies of Reading in Contemporary American Criticism*, ed. William V. Spanos, Daniel O'Hara, Paul Bové (Bloomington, Ind., 1982), 24–25; originally in *Boundary 2*, VII (1979), and revised and dulcified a bit in *The World, the Text, and the Critic* (Cambridge, Mass., 1983), 172.

(like Milton's Death "that shape had none"), too often irrationally taken to be inclusive, taken to be the nature of truth or the truth of nature, which is to say, too often *taken mythically.* Mumford's, Tate's, and Eiseley's oppositional writing is more the execution of a series of strategies to promote the recovery of sight to the blind than a political campaign to drive the money changers from the temple. But no reader whose vision has been confirmed or enhanced by their writings would willingly endure the money changers there.

Among the ideas and arguments that have enabled this study in more general ways than footnotes can easily indicate, an early and persistent one has been Louis Martz's case for a meditative tradition in literature, with Ignatius of Loyola as a sort of presumptive patron saint. Intending no flippancy, I imagine the saint to say, like Browning's bishop, "Nephews—sons mine." If not strictly meditative, the three essayists of this book are, I think, more akin than the three whom Jacques Barzun yoked in troika half a century ago as the progenitor party of revolutionary progress, in his *Darwin, Marx, Wagner.* But Barzun was looking back three-quarters of a century and could with clarity and propriety come nearer a summing-up. I hope to initiate further dialogue. At a crucial early stage, Herbert Schneidau's *Sacred Discontent: The Bible and Western Tradition* and, more recently, Fred G. See's *Desire and the Sign in American Literature,* which I read in typescript, have helped me immensely in recognizing some terms that have come to seem essential to my argument. And the chresmologic presence of Lewis Simpson has meant more than the annotation in what follows can suggest, as his countless admirers will surmise.

It is a pleasure to acknowledge the help of Maryellen Kaminsky, with the Eiseley Papers in the University of Pennsylvania Archives, and of friends and associates of Eiseley with whom I had the privilege of contact: Reuben Reina and Ginny Lathbury, in the anthropology department at the University of Pennsylvania, Jean Adelman, in the department library, Caroline Werkley, former editorial assistant to Eiseley (since retired to Moberly, Missouri), C. B. Schultz, of the Nebraska Academy of Science, and James Gulick, of the Heritage Room of the Lincoln Public Library, were forthcoming and informative, as were the staff at the Nebraska Archives. Helen Heinz Tate, Allen Tate's widow, has been collegially generous in correspondence and permissions. Janet Lewis has kindly granted me permission to quote from the correspondence of Yvor Winters. The staff of the Rare Book and Manuscript Collections at the Firestone Library, at Princeton University, make it a joy to work on the Tate Papers there; I was similarly assisted by librarians at the Library of Congress, the staff at the University of Virginia Library, with its more modest holdings, and that at the State University of New York at Buffalo Library, where I examined a few items. Very early, while I was still teaching at SUNY-Buffalo, Tom Evans and Dan Howell were exemplary research assistants. The departmental ambience at Buffalo and at LSU has helped me immeasurably: I have benefited from opportunities to teach Mumford and Eiseley at Buffalo, from a summer research grant awarded by the LSU Research Council, from a research travel grant made by the Southern Regional Education Board, and most important, from intellectually engaging and provocative colleagues. I am in their debt, although they may wish that I had drawn more amply on their wisdom. Claudia Scott has word-processed and reprocessed with wonderful cheerfulness. Herbert Schneidau has given valuable general advice on organization. Margaret Dalrymple and Catherine Landry, at Louisiana State University Press, have made valuable suggestions about presentation. Barry Blose, who edited the book at the Press, has made a multitude of suggestions in behalf of clarity—a number of which I have resisted, attributing some problems of understanding to the complexity of the material.

Joan Lambert Carrithers, dedicatee of an earlier book, has by countless reports over the years from her profession of social work alerted me to many particulars of the great cultural degeneration that Mumford, Tate, and Eiseley deplored. I see, too, with more than half a lifetime's perspective that my view of Eiseley and Mumford has been fostered by an older Gale, conservationist, engineer, and my father, and that my receptivity to Tate has been fostered by Hope, discerning student of words and the ways of their users, and my mother. This book is partly theirs, too. Finally, Sandy Carrithers and Mary and Christopher Snyder have spurred my care as they, now young adults, engage that world none too brave nor new: this book is dedicated to them.

ABBREVIATIONS

I. Lewis Mumford

CC   *The Culture of Cities*. New York, 1938. Vol. II of Mumford, *The Renewal of Life*. 4 vols.

CL   *The Conduct of Life*. New York, 1951. Vol. IV of Mumford, *The Renewal of Life*.

CM   *The Condition of Man*. New York, 1944. Vol. III of Mumford, *The Renewal of Life*.

F&K   *Findings and Keepings*. New York, 1975.

FJO   *The Letters of Lewis Mumford and Frederic J. Osborn: A Transatlantic Dialogue, 1938–1970*. Edited by Michael Hughes. New York, 1972.

SfL   *Sketches from Life: The Early Years*. New York, 1982.

T&HD   *Technics and Human Development*. New York, 1967. Vol. I of Mumford, *The Myth of the Machine*. 2 vols.

TC   *Technics and Civilization*. New York, 1934. Vol. 1 of Mumford, *The Renewal of Life*.

TPOP   *The Pentagon of Power*. New York, 1970. Vol. II of Mumford, *The Myth of the Machine*.

VWB   *The Van Wyck Brooks–Lewis Mumford Letters: The Record of a Literary Friendship, 1921–1963*. Edited by Robert E. Spiller. New York, 1970.

W&D   *My Works and Days: A Personal Chronicle*. New York, 1979.

II. Allen Tate

Col.   *Collected Essays*. Denver, 1959.

DD   *The Literary Correspondence of Donald Davidson and Allen Tate*. Edited by John Tyree Fain and Thomas Daniel Young. Athens, Ga., 1974.

For.   *The Forlorn Demon: Didactic and Critical Essays*. Chicago, 1953.
JPB   *The Republic of Letters in America: The Correspondence of John Peale Bishop and Allen Tate*. Edited by Thomas Daniel Young and John J. Hindle. Lexington, Ky., 1981.
Hov.   *The Hovering Fly, and Other Essays*. Cummington, Mass., 1948.
L   *The Lytle-Tate Letters: The Correspondence of Andrew Lytle and Allen Tate*. Edited by Thomas Daniel Young and Elizabeth Sarcone. Jackson, Miss., 1987.
Mem.   *Memoirs and Opinions, 1926–1974*. Chicago, 1975.
Reac.   *Reactionary Essays on Poetry and Ideas*. New York, 1936.
Reas.   *Reason in Madness*. New York, 1941.
Rev.   *The Poetry Reviews of Allen Tate, 1924–1944*. Edited by Ashley Brown and Frances Neel Cheney. Baton Rouge, 1983.

III. Loren Eiseley

All   *All the Strange Hours: An Excavation of a Life*. New York, 1975.
Ano.   *Another Kind of Autumn*. New York, 1977.
Dar.   *Darwin's Century: Evolution and the Men Who Discovered It*. Garden City, N.Y., 1958.
Fir.   *The Firmament of Time*. New York, 1960.
Imm.   *The Immense Journey*. New York, 1957.
Inno.   *The Innocent Assassins*. New York, 1973.
Inv.   *The Invisible Pyramid*. New York, 1970.
Man   *The Man Who Saw Through Time*. New York, 1973.
Mr.X   *Darwin and the Mysterious Mr. X: New Light on the Evolutionists*. New York, 1979.
Night   *The Night Country*. New York, 1971.
Notes   *Notes of an Alchemist*. New York, 1972.
Star   *The Star Thrower*. New York, 1978.
Unex.   *The Unexpected Universe*. New York, 1969.
Wings   *All the Night Wings*. New York, 1979.

MUMFORD,
TATE,
EISELEY

I

MUMFORD AND THE
RENEWAL OF LIFE ❧

*I would wager that the psychoanalyst would put the fre-
netic lover of technology and its enchanted detractor in the
same category. He would ask whether it is not the same
distortion of language and enjoyment which animates both
and delivers the first over to infantile projects of domination and the second to the fear of things he cannot control.*

—Paul Ricoeur, *The Conflict of Interpretations*

*Since the seventeenth century the trend in biology had been
to reduce living organisms to mechanisms; Usher, Mumford,
and Giedion all wanted to work in the opposite direction
and raise our conception of mechanisms as organisms or as
parts of a larger organism.*

—Arthur P. Molella, "Inventing the History of Invention"

What is happening to us, between us? What does
our situation in late-twentieth-century America have to do with the machine, broadly con-
ceived, the machine we have made but that threatens to make us in reductive ways? How shall we continue as makers rather than servants or
drones? What shall I, a twentieth-century American, do for, and as, my
life in this land between Europe and vanished frontier, between European history and North American prehistory, between heirship of the
ages and ominous, history-threatening present and future? How can we
with decent mutuality conceive and signify these concerns, these social
discoveries? Shall we not, with that, expose false names of a false god,
strike through the masks of Moloch? Shall we discern a new name of the
true god whose recognition is always in the psalmist's ascription a "new
song"? Some such conjugation of questions animated Lewis Mumford,
Allen Tate, and Loren Eiseley, and animates this study of the ways their

essayistic writings are confluent. (Indeed, the same questions impel my secondary attention to the ways the writings of the three men differ and my occasional references to what they write besides their essays.)

They were prophets, consciously and explicitly so; accordingly, this study is partly an essay in definition of the stance of prophecy. The three felt a sort of Old Testament calling to unmask false gods and denounce deathly practice and doctrine, in the name of—to put the matter generically—the true god of life. Moreover, prophecy meant for them, as Augustine put it in *De doctrina christiana,* to speak always of great things (*re magna*), whether grandly or humbly or in the middle style of self-subordination to the subject matter (*granditer, humile, submisse*). Their rhetorical behavior I will try to anatomize, while acknowledging the problems in doing so. Most obviously, the three wrote voluminously and in a great range of venues. Yet there has been very little consideration of the great bodies of work by them which are addressed here, and no consideration, to my knowledge, of all three of the writers, or even any two, together. In any case, their writings are not so familiar to some readers of this study as, say, *Macbeth, The Waste Land,* or *The Great Gatsby.*

It is their essays that I shall consider primarily, essays often gathered in book-length collections in the case of Tate and Eiseley, essays grandly conceived and executed at book length and ultimately clustered by Mumford. This might be another problem. Van Wyck Brooks wrote sympathetically to his longtime friend Mumford, "Curiously enough, *writing,* in this age, in this country, means simply fiction and poetry (and sometimes the drama) and writing here is a cult too while it excludes from the field all the Renans and the Carlyles and the Emersons of our day" (*VWB,* December 20, 1945). But the times they are a-changin': if the humanities wing of the academy has been slow to honor discursive prose in general and the essay in particular, the vanguards of literary theory are getting to figures from Montaigne to Henry Adams now, and—not to put too fine a point on it—the academy regularly accords something to its own practitioners of the scholarly essay.

But why these three? One might think (in no particular order) of Annie Dillard, Aldo Leopold, Bruno Bettelheim, J. R. Oppenheimer, Buckminster Fuller, B. F. Skinner, Nora Ephron, Lewis Thomas, George Will. The genre has flourished. A short answer, true but too vaguely general to be satisfactory, is that I have found each of these three peculiarly compelling, and found an intellectual confluence arising from such diverse personal histories to be singularly provocative. Moreover, or specifically, Mumford and Tate and Eiseley share an unusual degree of searching and constant mindfulness that *nature* and *science* and *self* are profoundly cultural ideas. Tate might have objected to the term *culture,* and Mumford surely would have, though not to all its implications. With regard to the problematic of culture, the three writers have responded by approaching from many sides the attendant question of what is now entailed in leading an American life. Their questioning has amounted to an extraordinarily profound challenge to the more or less consciously received religion—what Mumford came ultimately to call the "myth of the machine." They largely coincide in finding profane sacralism degrading to the past, perverse now, and baneful for the future. Each has his characteristic though by no means exclusive emphasis: Mumford, so to speak, on the liturgy or ontology of the false religion, Tate on its systematic theology or epistemology, Eiseley on its neutered Gnostic year, or eschatology. In varying degrees of eloquence and of dramatic dialectic they witness against the misplaced zeal without which a religion is not a faith.

Among American twentieth-century essayists, all of whom grumble, some of whom criticize semigenerally, and some of whom (Fuller, say, or Skinner) are so much proponents of a private and sinisterly immanent religion as to criticize by implication all else, these three stand out as the prophets simultaneously farthest in and farthest out—penetrating deepest into the beguiling *eidolon* of their culture, and distancing themselves in imaginative and existential rejection of it and the fevered semilife it engenders.

I shall examine Mumford first because in the thirties he was established in a way that even Allen Tate was not, and in a way that Eiseley was not until the fifties. He also seems initially more accessible. His obviously extensive view seems familiarly discursive from a recognizably cosmopolitan American standpoint in Western civilization, although as we shall see, puzzling eddies occur in his great flow of discourse. Tate's radically historical and intellectually intensive view requires some teasing-out, as do the implications of Eiseley's scientific breadth and stunning personal intensity. The historical sense, so central to all three men, and profounder with them than with most other American essayists, is of a more familiarly direct and explicit sort with Mumford than with either Tate or Eiseley. Mumford gives and glosses the phrase *the myth of the machine.* Finally, there is something more than a little Miltonic in the self-conscious lifelong preparer and impassioned addresser of giant prophetic essays to his people and their European neighbors, a prompter of the age to quit its clogs, a citizen who turned from exhilaration with the revolution to the sternly loving admonition that "new Presbyter is but old Bishop writ large."

That the prophetic promptings of the three are loving in diverse yet concordant ways seems to me as crucial as anything else. Tate quotes the biblical admonition "He that saith he is in the light, and hateth his brother, is in darkness even until now." In a darkness where mysteries are called problems and where ignorant armies of failed positivisms and instrumentalisms clash by night, these are three who would speak light to us.

In the "prefatory note" to *Technics and Civilization,* Mumford wrote that "the first draft of this book was written in 1930," but that "up to 1932 my purpose was to deal with the machine, the city, the region, the group, and the personality within a single volume." But, not unlike Prince Hal's Hotspur of the north, wanting action, he expanded the project into a tetralogy, *The Renewal of Life,* on which he worked over twenty years: *Technics and Civilization* (1934, 1963), *The Culture of Cities* (1938, 1970), *The Condition of Man* (1944), and *The Conduct of Life* (1951, 1970).'

If the drafts of 1930 and 1931 become available, they will invite comparison with the finished series, so extraordinarily grand in conception and execution. Any smile one might feel at Mumford's initial supposition, at age thirty-five, that he could do the project in one volume must fade in wonder that he could do anything of the sort in four while publishing six other books, several pamphlets, occasional reviews, and regular columns ("The Art Galleries" and "The Sky Line") in the *New*

1. Here, as elsewhere, I have used Elmer S. Newman's *Lewis Mumford: A Bibliography, 1914–1970* (New York, 1971). For some account of the expansion, see *SfL,* Chapters 31, 32.

Yorker. It was, in its scope and orientation, a utopian project from a man whose first book had been *The Story of Utopias* (1922, 1941, 1959, 1962). Moreover, the original scheme could not but respond to the enormous facts of World War II and the death on the Italian front in 1944 of the Mumfords' son, Geddes (the subject of *Green Memories;* 1947).

Technics and Civilization reads today as a remarkable production, and the more so for being written in the thirties, by an American, an American under forty. The youth of the writer might suggest itself even to a reader unaware that Mumford's birthyear was 1895. The exuberance at times may seem almost boyish—or naïve, of which more in a moment. He revealed in the now-published letters to his old friend Van Wyck Brooks not only ideas about the thing but the thing itself: "You . . . have not been showing, as I have, latent capacities for mob oratory" (*VWB,* April 12, 1935). And, closer to the very composition: "a book to out-Bentham the Benthamites, to out-Marx the Marxians, and, in general, to put almost anybody and everybody who has written about the machine or modern industrialism or the promise of the future into his or her place" (*VWB,* June 21, 1933).

Neither the mob oratory nor the excess is pervasive, nor are rivals put in their place without a judicious range of approbation or demurrer. Indeed another striking feature of the book is its appropriateness, no less than that of Eliot's *Waste Land,* for the most famous discarded title of the generation: "He Do the Police in Different Voices."

Brooks had written a decade earlier that "Emerson was right when he said that it only needs two or three men to give a new turn to the public mind" (*VWB,* September 13, 1925). Accordingly, it appears, Mumford wrote in *The Renewal of Life* and the subsequent trilogy comprising *The City in History* and the two volumes of *The Myth of the Machine* to those who might be presumed to formulate their civic attitudes and practices—judicial, administrative, commercial, aesthetic—on the basis of 450-page treatises. Insofar as such members of the community of mind and the republic of letters coincided with the congregation of authority, a Mumfordian volume of argument might expect to contend with any other discourse on even footing. Perhaps Mumford did not sufficiently attend in the thirties to the dark possibility that the public realm might be dominated by the merely venal or the slogan-obsessed, or the intellectually lazy. Nor, before television, did he anticipate domination of the public by simplistic visual images. He had heard Franklin D. Roosevelt talk like a "top university professor" (SfL, 478). Could he imagine an Eisenhower commanding staff assistants to "boil this down to one page"?

Partly Mumford was and remained the adult-education teacher, although Time's review of Technics and Civilization in the issue of May 7, 1934, called him a prophet. He taught "The Machine Age in America" in the Columbia University evening extension division from 1932 to 1935 (SfL, 448–49). In the book, he backgrounds his argumentative brief, and foregrounds large assemblages of information in broadly chronological array. In this, his tone is characteristically the exhibitor-polemicist moving confidently among primary and secondary texts in Latin, Italian, German, and French. It would have been by itself a considerable service to draw the attention he did to the writers he used and applauded: they include Werner Sombart, especially for Der moderne Kapitalismus (still untranslated, although single-volume works by Sombart have appeared in English); Frédéric Le Play, whose six-volume Les Ouvriers européens of 1879 Mumford called a landmark needing follow-up; Fritz Schumacher, whom he ranked superior to the more modish Oswald Spengler; and Thorstein Veblen, not forgotten in 1934 but not, as far as I know, elsewhere paired with Sombart as "after Marx . . . perhaps the foremost sociological economist." He used Max Weber for the "romanticism of numbers," a concept that would grow on his darkening horizon, and praised R. H. Tawney, in one of his annotations scattered in the bibliography, as "an able economist and a humane mind." He cited as "invaluable" Lynn Thorndike's History of Magic and Science and acknowledged as mentors Victor Branford and Patrick Geddes, giving Geddes' term ecology, as well as the idea it represented, one of its first serious treatments in America.

But it would be as inadequate to think of Mumford as merely reprising other writers' texts as it would be to think of those texts which were rich in polemics and interpretation as "secondary." They were in a sense primary. As his acknowledgments, illustrations, and memoirs attest, he benefited equally from museums, especially the Museum of Natural Sciences and Technics in Munich (SfL, 467). In addition, indeed more fundamentally, he was a walker, student, artist, and trustful and feeling inhabitant of New York City most of his first thirty years. The Mumfords had spent a summer on Martha's Vineyard in 1927, and the time after autumn, 1929, until late spring, 1933, just before he "planned to start writing *Technics and Civilization*," they had passed at their early-nineteenth-century house in tiny Leedsville, New York, which was gradually taking "possession" of them (*SfL*, 466, 452, 484). As any persistent reader of Mumford surely notices, and as he disarmingly quotes an extramarital lover to have written him just earlier, he had a "genius for establishing direct firsthand authentic contact with physical and mental reality" (*SfL*, 464). One imagines Mumford receiving such words, so delivered, with a kind of prayer that it be ever so, or become so. In any case, the point will bear returning to, repeatedly.[2]

The strong argumentative spinal column of *Technics and Civilization* makes of its adult-education substance something of a hortatory structure tending toward the sermon at moments and more frequently toward apologetic, but with several internal conflicts to be noted in a moment. The strength and interest of the core argument lie in its attempt to interrelate infant science, "the machine," and the Industrial Revolution as they bear on life both existentially and temporally-historically at large.

The matter has larger resonances. Andreas Huyssen has argued forcefully in his recent *After the Great Divide: Modernism, Mass Culture, Postmodernism* that since Henri, Duc de Saint-Simon, in the early nineteenth century, "the avantgardes of Europe had been characterized by a precarious balance of art and politics, but since the 1930s the cultural and political avantgardes have gone their separate ways," and he is generally persuasive. But the Mumford of *Technics and Civilization* and *The Culture of Cities*, the Mumford who referred to Saint-Simon in text and bibliography only in *The Condition of Man* (1944), *was* articulating what Huyssen usefully calls the "hidden dialectic: avantgarde–technology–mass culture." He *was* calling into question the "Marxian separation of economy and culture as base and superstructure," *was* insisting that an "increasingly technologized life world" has not only transformed everyday life but "has also radically transformed art." It is

2. In one of the most sweeping and suggestive articles on Mumford, Christopher Lasch notes that his genius for "concrete historical problems" enriches *Technics and Civilization* and *The Culture of Cities* and that the neglect of such problems correspondingly impoverishes *The Condition of Man* and *The Conduct of Life*. See Lasch's "Mumford and the Myth of the Machine," *Salmagundi*, Summer, 1980, pp. 11–12.

not in Walter Benjamin's "work of the 1930s that the hidden dialectic between avantgarde art and the utopian hope for an emancipatory mass culture can be grasped alive for the last time."[3] It cries from the first two of Mumford's books considered here and is animated largely, I think, by a notion I shall examine—a notion that romantically-artistically reimagines and reassimilates mechanism as part of a "larger organism," in Arthur P. Molella's words. That reconceiving changes tonality in *The Condition of Man* and *The Conduct of Life,* to reemerge in the service of a prophetic strain in *The Myth of the Machine,* as we shall see. The resulting complex organicized conception of mechanism, including ambivalences and inconsistencies akin to some that Huyssen teases out of Theodor Adorno, energized Mumford's writings well before the early 1970s, when Adorno and Max Horkheimer commented, in a passage that Huyssen translates, to the effect that persons are victimized and objectified in part willingly by seductive technological forces.[4] But my focus is not the degree to which Mumford is an American Adorno or a homegrown Horkheimer; rather it is, in the words of the title he gave a scrapbook of autobiographical materials, his "findings and keepings" in his long trek through American culture.

Mumford characterizes infant science as eliminating allegedly "secondary" qualities in favor of the measurable, as neutralizing the observer with regard to measurement, and as isolating fields of investigation by limitation or subdivision. (That characterization became widely current only somewhat later, with his help, it would seem.) He identifies "the machine," accordingly, as "the entire technological complex . . . the knowledge and skills and arts derived from industry or implicated in the new technics . . . various forms of tool, instrument, apparatus and utility as well as machines proper" (*TC,* 46, 12). Science and the machine—by this definition, made for each other—are consonant with capitalism by the "same kind of power: the power of abstraction, measurement, quantification" (*TC,* 25).

We are apt to think of the vast Western gravitation toward abstraction during recent centuries as fostered most radically by literacy, by writing as a "technology which alters thought," and by "print culture." Walter Ong, Elizabeth Eisenstein, Marshall McLuhan, and others have brought us to that conviction but have done so largely in the past

3. Andreas Huyssen, *After the Great Divide: Modernism, Mass Culture, Postmodernism* (Bloomington, Ind., 1987), 4–6, 21, 9, 14.
4. *Ibid.,* 26.

quarter century.[5] Scant wonder that Mumford rather slights printing and language; it is evidence of his "authentic contact with . . . mental reality" that he posits Protestantism as resting in practice "firmly on the abstractions of print and money" (*TC*, 43).[6] He does not discount print or language,[7] but in this early book he relatively neglects both to emphasize the psychological reward of incrementally getting control, by abstraction, over adversive natural and social scenes. He also underscores the abstractive tendency of reinforced *time* consciousness: "Space and time, like language itself, are works of art, and like language, they help condition and direct practical action. Long before Kant announced that time and space were categories of the mind . . . mankind at large had acted on this premise" (*TC*, 18). Of greater interest for Mumford's purposes is his more particular point that Western Europe produced the momentous development of an abstractive-time alternative to organic time—"in the seventh century, by a bull of Pope Sabinianus . . . that the bells of the monastery be rung seven times in the twenty-four hours," and in the mid–fourteenth century, by the commonplace division of hours into minutes and seconds by a "well-designed 'modern' clock . . . at Paris." It was the newly common clock, "not the steam-engine," that was the "key-machine of the modern industrial age," in that it "helped create the belief in an independent world of mathematically measurable sequences" (*TC*, 13–15).

After Mumford read Kant but apparently before he read Ernst Cassirer or Suzanne Langer, the latter of whom he acknowledges in *The Condition of Man,* he was construing human life-in-culture in terms of symbolic forms. Whether language is for him the most fundamental of symbolic forms, whether it is somehow more essentially an outward sign of some deep inner structure of the mind than space, time, or a work of, say, musical art, he does not seem to affirm clearly enough yet that his answer would date his position.[8]

5. For magisterial summary and bibliography, see Walter Ong, *Orality and Literacy: The Technologizing of the Word* (New York, 1982); for brilliant exposition, substantiation, and bibliography, see Elizabeth Eisenstein, *The Printing Press as an Agent of Change in Early Modern Europe* (New York, 1979).

6. For the history of a Protestant attitude toward interest (acceptable at moderate levels) and a Roman Catholic attitude (damnable usury at any level), see Herbert Luethy, "Once Again: Calvinism and Capitalism," *Encounter,* XXII (1964), 26–38.

7. When he writes of the breaking of the "tyranny of the written text" by alchemy and painting (*TC,* 42), the target is not printing but prelacy.

8. For a persuasive argument that some very early carved markings are lunar calen-

In any case, he was consistent after the early 1930s in insisting that abstractive, taxonomic time differs critically from *history*, which he wants understood as a thick web of never completely ascertainable influences, choices, consequences, and continuing options, and that it differs just as critically—he bows to Bergson—from experienced duration. These points, not without complexity, have perhaps been too familiar since midcentury to require much notice, but they are profound enough. Mumford helped win them currency, and a host of subsequent works narrower in focus than *The Renewal of Life* and *The Myth of the Machine* read like footnotes to his argument about the thickness of history, the significance of experienced temporality for the individual, the human tendency to mythicize reductively, and the dehumanizing and seductive quality of mechanism.

One of the most fundamental efforts of the five hundred pages of *Technics and Civilization,* as of the later volumes in *The Renewal of Life,* as well as *The City in History* and *The Myth of the Machine,* is toward the thickening of history. History, for Mumford, is the cultural ground of the present, remaining, very often at least, an optional active part of the present and furnishing landmarks for the evaluation of challenges in it. His attempt is, in part, to situate technics in the realms of historical and philosophical discourses. Technics is the "result of human choices and aptitudes and strivings, deliberate as well as unconscious, often irrational when apparently they are most objective and scientific. . . . No matter how completely technics relies upon the objective procedures of the sciences, it does not form an independent system, like the universe: it exists as an element in human culture and it promises well or ill as the social groups that exploit it promise well or ill. . . . It is the human spirit that makes demands and keeps promises" (*TC,* 6).[9]

There, at the beginning, is a variation on a theme that will recur, that of history and culture as the ground of technics—of technics as a connection between historical culture and individual or group desire, even

---

dars, that is, constitute evidence of "time factoring," with abstractive observation and representation, see Alexander Marshak, *The Roots of Civilization: The Cognitive Beginnings of Man's First Art, Symbol, and Notation* (New York, 1972).

9. I cannot determine whether he had any acquaintance with Gabriel Marcel's idea of man as *essentially* the maker (and sometimes the keeper) of promises. It looks like an independent anticipation of expressions in Marcel's *Du réfus à l'invocation* (Paris, 1940), which has appeared in English as *Creative Fidelity* (New York, 1967).

between historical culture and the individual body. Since the individual human body is for Mumford one of the two means and ends to which it is necessary to refer in evaluating any proposal of politics or any production of technics, it is polemically and historically significant that he goes on to become one of the first in America to identify a displacement of bodily shapes, movements, and affections upon the machine, and to decry the concurrent rise of "a new religion . . . a new Messiah: the machine" (*TC*, 55, 45).

Less persuasively—indeed, writing as if the point were self-evident and needed no argument—he explains the rise of the new allegiance and new religion in the tendentiously metaphorical terms he deems suitable to the late Roman Empire ("putrefaction and decay"; *TC*, 44) and to medieval Christianity ("a dream," "a lifting mist"; *TC*, 28, 31): "A live machine was better than a dead organism; and the organism of medieval culture was dead" (*TC*, 44). He almost out-Gibbons Gibbon, who described the decline of the Roman Empire as the triumph of barbarism and religion.

A tack of Mumford's that is more useful for explication, even if little more decisive with regard to cause, is his placing of emphasis on control: "In attempting to seize power [over nature] man tended to reduce himself to an abstraction, or, which comes to almost the same thing, to eliminate every part of himself except that which was bent on seizing power" (*TC*, 31). The machine, he notes, gained importance most slowly in the "life-maintaining" context of agriculture, whereas it "prospered . . . in the monastery, in the mine, on the battlefield" (*TC*, 36). The book's argument as a whole tends to conceive these places, in their institutional nature, as parts or even species of what by the 1960s he was calling megamachines.

Yet the Mumford of the thirties was not so adept as the post–World War II Mumford in explicitly identifying the problematical attitudes persisting in quite various institutional contexts. Nevertheless, one is struck by the interpretive observation that could serve for the second-order interpretation: "To assume that a later point in time *necessarily* carries a greater accumulation of values is to forget the recurrent facts of barbarism and degradation" (*TC*, 184); "Right down to the World War an unwillingness to avail itself of scientific knowledge or to promote scientific research characterized paleotechnic industry throughout the world. Perhaps the only large exception to this, the German dye industry, was due to its close connection with the poisons and explosives necessary for military purposes" (*TC,* 194). Those remarks in Chapter Four, "The Paleotechnic Phase," conclude a canvass of socioenvironmental horrors attending early "carboniferous capitalism," and an indictment of the self-serving conservatism, early and late, of paleotechnic institutions. At this stage in his career as cultural critic, he draws no overt analogy between the idolatrized medieval conservative ecclesia and paleotechnic industry, and no more than a guarded, implicit analogy between the German dye industry and enterprising Benedictine monastic rule.

He more than observes, he proclaims, that the bad, the false, and the ugly from a past of closely associated mining and militarism still impinge upon us, as do some of the worst effects of carboniferous capitalism. Like Marx, he may be faulted for underestimating the self-reforming efforts of capitalist societies with Western-style governments and labor movements. But he differentiates, in terms of the effects on bodily vitality and, what is often more ambiguous, on cultural vitality, the good from the bad elements of the "eotechnic phase," characterized by handwork, water, wind, and wood, and the good from the bad developments and transformations of the "neotechnic phase," which is primarily electricity-related.

From *The Golden Day* (1926) onward, Mumford has tended to use the word *culture* when he means primarily the more or less shared meanings, values, and aspirations of a society, the word *civilization* when he means primarily outward and visible or audible signs of a society's culture. Civilization to culture is virtually a special case of signifier to signified. His career has been an extensive and deepening reading of the cultural implications of artifacts, of the implications of artifacts for other elements of civilization, of culture's expressions in its omnipresent artifact (language), and of discrepancies between artifact and artifact, artifact and value, and value and value. In *Technics and Civilization,* he scans both civilization and culture to discern continuations and variations of that which has been considered the good, which alternate between "good for the social personality" and "good for the human body," occasionally encompassing both. Nowhere in *The Renewal of Life* does he mention Ferdinand de Saussure, but he continually scrutinizes the relationships of signifier and signified in a manner as if Saussurean, yet with a generous appreciation for "meaningful understanding" as *embodied:* Mumford's signifiers are not altogether arbitrary—at least they should not be.

After a brief on the intimate relationships of mining, capitalism, and militarism in the fifteenth and sixteenth centuries—militarism indeed fostering the earliest mass production, rivaled only by printing (*TC*, 90)—he even comments on the social with a bodily metaphor: "It was unfortunate for society at large that a power-organization like the army, rather than the more humane and cooperative craft guild, presided over the birth of the modern forms of the machine" (*TC*, 96). Most generally, most fundamentally, he privileges creation over production:

> Creative activity is finally the only important business of mankind, the chief justification and the most durable fruit of its sojourn on the planet. The essential task of all sound economic activity is to produce a state in which creation will be a common fact in all experience: in which no group will be denied, by reason of toil or deficient education, their share in the cultural life of the community, up to the limits of their personal capacity. Unless we socialize creation, unless we make production subservient to education, a mechanized system of production, however efficient, will only harden into a servile byzantine formality, enriched by bread and circuses. (*TC*, 410)

The evil that opposes this, plainly enough, is the mechanical working pattern in asocial contexts—mindlessly repetitive tasks on the assembly line in a penalty-oriented organization, say. The socialized creativity Mumford advocates for the workplace would seem to be more or less jointly animated by two spirits, that of comradeship and that of marginal freedom among rules not malign that characterizes play and some gaming. It should aim, as part of its "assimilation of the machine," to establish a closeness between design and production or operation which is analogous to that of handicrafts (*TC*, 351), in short, feedback.

Mumford was prescient, remarkably so for one writing when the persisting depression, and workmen's gratitude to have jobs at all, were masking the costs—costs to quality, morale, attendance, sobriety, and the like—attendant on a reductively, shortsightedly "efficient" operation of assembly lines and such. And one is bemused by the mixed commentary on Mumford's point that a half-century of experience provides: the by all accounts unforgettable striving and mutually competitive emulation by the committees in the multicommittee team J. Robert Oppenheimer organized at Los Alamos to develop the atomic bomb, the more recent efforts of Volvo executives to organize their auto production as multiple miniteam efforts, the American exhortations to team spirit and quality (sometimes with a bow toward supposed Japanese models), the employment of computerization to diversify mass production (as for local advertisements in national magazines or special-order features or combinations on automobiles). Although it may be too early to say, the socializing of creativity that is occurring in the diversification of mass production appears to involve a few computer artists and increasingly automated production. Mumford was presumably not surprised, having exhibited in 1934 a foresight reaching beyond mere technological unemployment to a sense of what Norbert Wiener in the fifties was to call, in *Cybernetics* and *The Human Use of Human Beings*, a second industrial revolution, a revolution obviating most mental drudge work as the first had obviated much physical drudge work (*TC*, 228, 278–80; see also p. 101 about early playful motifs).

Mumford's touch was not always so sure. Opposing the short-range, narrowly conceived productivity figures and cost accounting that still require and are today starting to receive opposition, he could write: "The permanent gain . . . is in the relatively non-material elements in culture . . . or directly in life itself. . . . As John Ruskin put it, *There is no Wealth but Life;* and what we call wealth is in fact wealth only when it is a sign of potential or actual vitality. . . . The important factors here are not quantities but ratios . . . of mechanical effort to social and cultural results" (*TC*, 377–78). How overstated and precarious that *permanent* looks from the viewpoint of the later Mumford, not to mention his University of Pennsylvania colleague Eiseley. How thin a mask *ratios* is for the continuing positivistic quantification of efforts and results that the word condones. Mumford's writings of the thirties would be superseded, even dead, if this sort of thing were characteristic.

But arguably it is not. In being somewhat more particular, Mumford can be more fruitfully provocative. He writes of the eotechnic period in northwestern Europe: "In every department of activity there was equilibrium between the static and the dynamic, between the rural and the urban, between the vital and the mechanical. . . . Many of its artifacts are still in use. . . . Far from moiling day and night to achieve as much as it did, it enjoyed in Catholic countries about a hundred complete holidays a year" (*TC*, 148). The balance in the workday-holiday articulation of life and the balances within the other polarities he identified (apart from the other issue of their validity for the period and region in question) suggest in the representation *equilibrium* a dynamic tension that Tate might have applauded. Or the idea of equilibrium might lead to an image of life as an alternation of systole and diastole. But Mumford does not use the image, and it is not clear how far the equilibriums he speaks of extend beyond his eye into the lives of those he so grandly surveys. Apart, perhaps, from the equilibrium between rural and urban, there seems in the context of his argument to be a case made for eotechnic life as well articulated for the arousal and gratification of a wide spectrum of desires. Mumford himself, of course, was alternately writing and gardening in Amenia, New York, as he formulated the argument, and was shuttling between the country and New York City.

With regard to the practical articulation of life and its appearances in the best of the eotechnic phase, before carboniferous capitalism, and in the best of neotechnic life since, he makes two points related to equilibrium. Capitalism and technics are, he says, a "solvent" of class and caste snobbery, and "thus at first important liberators of life" (TC, 323, 354–55). And, he asserts, "there is an esthetic of units and series, as well as an esthetic of the unique and the non-repeatable" (TC, 334). The second point, in behalf of repetitious machine forms and structural forms—efficient in design and production, and functional in operation—connects the new with the old in more ways than Mumford seems to have recognized in the thirties: not only, obviously, with antique geometrical design and classical architectural forms but also with the oral formulas and repetitiousness dictated by any oral society's economy of memory and any disordered society's desire for order. (He notes the multiple rewards of similar *days* under the Benedictine rule.)

The more fundamental connections he makes involve distinctions of true and false religion. I have already noted his concern with "mechanization and regimentation . . . projected and embodied in organized forms which dominate every aspect of our existence": European society, "by an inner accommodation, surrendered to the machine" (TC, 4). "Only as a religion can one explain the compulsive nature of the urge" (TC, 365; similarly TC, 169, 199, 291).

Christianity, likewise, seems—apart from some of its artifacts—to have impressed the Mumford of the midthirties as a religion of nonlife, institutionally and probably even doctrinally in league with the forces of repression: he speaks of synchretic "myths and offices that became Christianity" (TC, 107) and of "mythological constructions" (TC, 315). None of that is to be confused with the somewhat later demythologizing concerns of Judeo-Christian theology. But Mumford is not greatly concerned with Christianity, though not because he considers technics to be apart from it. Everywhere in his writings he makes clear that technics pervades society and is implicated in religious questions. The point is rather that *Technics and Civilization,* as propaedeutic-apologetic, addresses "some 'central' hypothetical audience, which . . . *should* exist" but which *The Renewal of Life* might have to create—as Van Wyck Brooks wrote to him (*VWB,* October 25, 1931).

Mumford appears to conceive of factory workers as a main repository of Chistianity but not as a considerable component of the central audience: "[Factory] workers themselves were thrown back upon the rubbish of earlier cultures, lingering in tradition and memory, and they clung to superstitious forms of religion which kept them in a state of emotional tutelage to the very forces that were exploiting them, or else they forfeited altogether the powerful emotional and moral stimulus that a genuine religion contributes to life" (*TC,* 408). In the larger context, his comment may escape superciliousness because he acknowledges that "*we* continue to worship the twin deities, Mammon and Moloch, to say nothing of more abysmally savage tribal gods" (*TC,* 264; my emphasis). Obviously such imagery, as when he devotes a subsection to "Mars and Venus" (*TC,* 96–101) exactly demythologizes and reorients what may be an orientation of some individual's or group's ultimate trust. He himself has appreciative, though hardly worshipful, words to say for Venus in that subsection and later when appraising the implications and consequences of neotechnic economy. In the latter passage, indeed, he notes contemporary advances in contraception and approvingly predicts what has come to be called, if extravagantly, the sexual revolution.

What religions are matters of more anxious concern? We have seen that a paganism of "the machine" and a Christian ecclesia that Mumford supposes to be allied with the mechanistic-commercial power structure are easy for him to dismiss. It might be argued that the combination of his startlingly high regard for hydroelectric plants and his petulantly low regard for toothpaste and household vacuum cleaners betokens tensions not fully conscious.[10] But his more unmistakably anxious address—as of an early church father in apologetic for the true faith and in polemic against insidious and tempting heresies—opposes communism and positivism with what might be called (in an almost

10. On hydroelectric facilities, see *TC,* Illustrations XII, 2; XV, 1. On vacuum cleaners, see *TC,* 277. The point is substantially repeated in *CM,* 88, and *T&HD,* 161.

contemporaneous phrase of José Ortega y Gasset's) a "few drops of phenomenology," but it does that inconsistently.

The romance of American intellectuals in the thirties with Marxism and the Soviet "experiment" is a familiar story. It continued for some even through the underreported policy of starvation in the Ukraine, through the Moscow purge trials, and up to the Russo-German Nonaggression Pact. Significant for our purposes are the indications Mumford gives of his idiosyncratic independence.

Technics and Civilization praises *Das Kapital* as full of "historic intuitions" (p. 216), as a "classic work whose historic documentation, sociological insight, and honest passion outweigh the defects of its abstract economic analysis" (p. 464). But "its description of price and value remains as prescientific as Ricardo's. The abstractions of economics, instead of being isolates and derivations of reality, were in fact mythological constructions" (*TC*, 216). And Marx, though right about the degradation of the worker, was wrong, Mumford thinks, in underrating the significance of utensils and utilities and in supposing that the middle class, small industry, and nationalism would all wane instead of wax (*TC*, 234, 189). On the other hand, he complains that capitalism has been in and of itself an amoral system of power for consumption and profit, with no "theory of non-profit-making enterprises and nonconsumable goods" (*TC*, 377, and *passim*). One could object that this is like Aquinas arguing that the intellect by itself never errs: Quite possibly true, but when has the intellect or capitalism ever been so isolated?

But Mumford's focus in these remarks on Marx and capitalism is on paleotechnic capitalism and industry of the nineteenth century, with their minimum of science, which gradually yielded, as the twentieth century arrived, to a closer alliance with science (as in electrification and attendant nonferrous metallurgy). And he was writing when the network of regulatory agencies, "safety net" agencies, and endowments for arts and humanities were far less extensive than has come to be the case.

What he proposes in the subsection "Basic Communism" in the final chapter of the book credits Edward Bellamy with first seriously proposing the guaranteed minimum income, albeit "in a somewhat arbitrary form," in his *Looking Backward*. Mumford requires quotation at some length:

> To make the worker's share in production the sole basis for his claim to a
> livelihood—as was done even by Marx in the labor theory of value he took

over from Adam Smith—is, as power-production approaches perfection, to cut the ground from under his feet. In actuality, the claim to a livelihood rests upon the fact that, like the child in a family, one is a member of a community: the energy, the technical knowledge, the social heritage of a community belongs equally to every member of it, since in the large the individual contributions and differences are completely insignificant.

[The classic name for such a universal system of distributing the essential means of life—as described by Plato and More long before Owen and Marx—is communism, and I have retained it here. But let me emphasize that this communism is necessarily post-Marxian, for the acts and values upon which it is based are no longer the paleotechnic ones upon which Marx founded his policies and programs. Hence communism, as used here, does not imply the particular nineteenth century ideology, the messianic absolutism, and the narrowly militarist tactics to which the official communist parties usually cling, nor does it imply a slavish imitation of the political methods and social institutions of Soviet Russia, however admirable soviet courage and discipline may be.] (TC, 403; Mumford's square brackets for a passage he italicized) [11]

What follows? A *scientific* "economy of needs" (TC, 412), rather than a Ricardian or Marxian mythological economy of price and value (TC, 216) or a capitalist economy of acquisition? Yes, a sort of biological positivism: "With the neotechnic phase . . . the scientific method, whose chief advances had been in mathematics, took possession of other domains of experience: the living organism and human society. . . . Physiology became for the nineteenth century what mechanics had been for the seventeenth; . . . organisms began to form a pattern for mechanism. . . . Similarly, the study of human life and society profited by the same impulses toward *order* and *clarity*" (TC, 216; my emphasis). One notices the predisposition toward taxonomic-visual stasis, or at least reductionism. In the assertion that the economic abstractions

11. In April, 1932, he had composed a "manifesto" with Edmund Wilson, Waldo Frank, John Dos Passos, and Sherwood Anderson, calling for the "ruling castes" to be "expelled from their present position" and a "temporary dictatorship of the class-conscious workers" to be set up (Edmund Wilson to Theodore Dreiser, May 2, 1932). On September 14, 1932, Mumford described himself to Van Wyck Brooks as "certainly post-Marxian" (VWB). By November 3, 1939, and again on February 10, 1940, he was expressing to Brooks guilt "for having remained so long indifferent to the fate of Communism in Russia, and so silent about the villainies of its dictatorship." Lasch interestingly aligns Mumford with American anti-Marxists and European antipositivist Marxists such as Antonio Gramsci and the Frankfurt school ("Mumford and the Myth of the Machine," 5–7).

of Ricardo and Marx were "mythological," rather than "isolates and derivatives of reality," Mumford can be seen to be wrestling with the perennially critical problem of *distance*—of what the optimum degree of spatial or conceptual *mediation* is between myth or deadly abstraction, on the one hand, and *immediate* engulfment, on the other. More broadly, Mumford is, like the biblical prophets but unlike contemporary cultural critics and social engineers, a utopian—aiming for a transcendently "acceptable year of the Lord." Else why bother? [12]

Part of Mumford lusts for "clarity," decisiveness, inclusiveness, and closure. He seems not to doubt that there is a *calculus* of life when he complains that even in neotechnics "the calculus of energies still takes precedence over the calculus of life" (*TC*, 249). No conscious irony shadows his pronouncement that the shrinkage of the machine to the provinces where its services are unique and indispensable is a *necessary consequence* of our better understanding of the machine itself and the world in which it functions" (*TC*, 258; my emphasis). It is difficult to know the extent of Mumford's biophysiological knowledge at the time and beyond my competence to cartograph the limited biophysiological knowledge of the time, within which one suspects Mumford's command extended little beyond important matters plain to an intelligent, interested layman.

What can readily be seen is a fervent trust in, or at least an eagerness to trust in, totalizing possibilities of a positivistic sort: "Perhaps Russia alone at present has the necessary framework for this planning [of production]" (*TC*, 389). Or: "When we begin to rationalize industry organically, that is to say, with reference to the *entire* social situation, and with reference to the worker himself in *all* his biological capacities . . . (*TC*, 413; my emphasis). Or even: "Good and bad, beauty and ugliness, are determined, not merely by their respective natures *but by* the quantity *one may assign* to them in any particular situation. To think closely with respect to quantities is to think more accurately about the essential nature and . . . functions of things" (*TC*, 328; my emphasis). This is followed by a medicinal example and a discounting of "pedantic . . . mathematical means." The hint of metaphorical proportion in the *but by* and in the arresting *one may assign* lift this out of sterile simplicity.

Mumford seems to want to believe, seems to believe at some moments, that such quantities might be found, sufficiently uniform, and as-

12. I have been moved to this conviction in argument with my colleague James Hardy.

signed by beings themselves sufficiently uniform. At such moments, he seems to think that dehistoricized featherless bipeds are responsible for discovery and invention, defined as consisting "like every other form of activity . . . in the interaction of an organism with its environment" (*TC*, 138). They, too, seem to be needed for devising an "arranged and purely artificial world language for pragmatic and scientific uses," as against the "cultural language for local communication" (*TC*, 294). Mathematics is indeed international, but the UN features simultaneous translation, English is the international language of flight control, and Esperanto seems passé.

Mumford's impulse to order is scarcely indecent, his energetic confidence far from unhealthy, his advocacy of remarkably inclusive planning not so megalomanic, ungenerous, or naïve as might appear to a late-twentieth-century eye:

> The technique of creating a neutral world of fact as distinguished from the raw data of immediate experience was the great general contribution of modern analytic science. This contribution was possibly second only to the development of our original language concepts, which built up and identified, with the aid of a common symbol, such as tree or man, the thousand confused and partial aspects of trees and men that occur in direct experience. Behind this technique, however, stands a special collective morality: a rational confidence in the work of other men, a loyalty to the reports of the senses, whether one likes them or not, a willingness to accept a competent and unbiased interpretation of the results. This recourse to a neutral judge and to a constructed body of law was a belated development in thought comparable to that which took place in morality when the blind conflicts between biassed men were replaced by the civil processes of justice. (*TC*, 361)

On the other hand, the late-twentieth-century mind may well wonder what a truly "neutral world of fact" could be and what relevance it might have to us if it were possible or conceivable. We may wonder whether data can ever be truly raw and whether experience is not always already mediated at least for postinfantile, that is, languaged, humans. We may wonder whether so tendentiously metaphorical ("built up") and nominalist a notion of language as Mumford's squares with the complexity of creativity in culture and whether the reports of the senses are so valid and reliable that they warrant such juridical prescription as Mumford seems to imply. It is one thing to say, "Thou shalt not dump phenol in the Mississippi River," quite another to say, "Thou shalt be a machinist, in Birmingham, for thy own good."

Besides, to compare applied science with jurisprudence—apart from trading on a possible confusion of *unbiased* with *disinterested*—is to invite awareness both of the cultural importance of applied science and of its conflicted and problematical situation. Anyway, sensory experience, for all that Old English maxims might proclaim beyond cavil that fire burns and frost freezes, tends to slide into the idiosyncratic or into metaphor or both. Mumford adducing a healthy new kind of personality in the age of the movie camera ("the matter-of-fact soul, naked, exposed to the sun on the beach") clearly reads like the private, personal, if not idiosyncratic, man who admired his wife and small son at the beach during their summer on Martha's Vineyard (see *SfL*, photograph after p. 278; *TC*, 244).

The expressions "rational *confidence*," "loyalty to the reports of the senses [of whomever?]," and "willingness" all reintroduce the issue of distance as much as they reflect any question of volition. They reach to the Other's mode of presence to myself, some mediated and moderate degree of intimacy. A pervasive and significant feature of Mumford's exposition is the recognition that machine forms can function in a manner parallel to that of the Other, if in a less richly extended way:

> Finally, the perfected forms begin to hold human interest even apart from their practical performances: they tend to produce that inner composure and equilibrium, that sense of balance between the inner impulse and the outer environment, which is one of the marks of a work of art. The machines, even when they are not works of art, underlie our art—that is, our organized perceptions and feelings—in the way that Nature underlies them, extending the basis upon which we operate and confirming our own impulse to order. The economic: the objective: the collective . . . [deemed] organic—these are the marks, already discernible, of our assimilation of the machine not merely as an instrument of practical action but as a valuable mode of life. (*TC*, 356)

Here again, the comparison provokes reflection and question. *Do* the artifacts of technology "underlie . . . our organized perceptions and feelings—*in the way that* Nature underlies them"? And if so, as Eiseley was to ask a generation later, how natural is nature? Introspection may disclose a strong "impulse to order," or not, but also varying degrees of confidence in its achievement. And would even a widely and confidently shared impulse to order necessarily be associated with some conception of the organic?

One other, briefer remark by Mumford must, I think, be cited in this connection: "It was, indeed, by the success of science in the realm of the inorganic that we have acquired whatever belief we may legitimately entertain in the possibility of achieving similar understanding and control in the vastly more complex domain of life" (*TC*, 327). *What* degree of belief is of course one of the central questions, and to "entertain" it rather than, say, hold it is to be mindful of mediation that distances presence. Will the sought understanding be similar in the way of one geometrical figure to another? in the way of equivalence? of analogy? of allegory? The prospect of "control" invites concern from any reader in this age of torture, concern with regard to custodians who might have no one to watch them. Does the acknowledgment of "vastly more" complexity in the "domain of life" not raise questions of whether differing orders of magnitude are involved and, indeed, whether differences of such degree do not become differences in kind?

My attention to these last few quotations has reversed the order they have in Mumford's book. His final chapter, "The Assimilation of the Machine," moves quasi-triumphantly to a concluding section, "The Objective Personality," and a flatulent "O altitudo" in the main text's final sentence: "Nothing is impossible." But I can argue that my constructed sequence is not grossly unfaithful to either the historical Mumford's order or the Mumford likely to prove engaging and rewarding to many of us at this stage in our shared American history. Certainly the passionately, learnedly historical Mumford would not want to transcend history (even if his secret sharer might). That Mumford can insist, "One begins with life; and one knows life, not as a fact in the raw, but only as one is conscious of human society and uses the tools and instruments society has developed through history—words, symbols, grammar, logic, in short, the whole technique of communication and funded experience" (*TC*, 370).

On the level of grand strategy, historical awareness does not always much perspectivize the diagrams of Mumford. In the conflict of endorsing Werner Sombart's interpretation that since 1914, capitalist modes of existence have been increasingly impregnated with "normative ideas," he writes that the indicated governance, by the state, of "banking functions" (in a large sense including capital investments, factories) "naturally . . . means a revolution" (*TC*, 421–22). But new-order planning "will *leave a place* for irrational and instinctive and traditional elements in society which were flouted, to their own ultimate peril, by the narrow forms of rationalism that prevailed during the past century" (*TC*, 417; my emphasis). What planner's formula yields a house of labor or house of intellect or house of family "*almost* big enough for my imagination?" asked a colleague of mine, settling into a new house. His words suggest how idiosyncrasy looms larger than *Technics and Civilization* allows but suggest also the very kind of ego strength that enabled Mumford in that book and later to leave, if not exactly a place, at least a possibility, for ranges of free human horizons and satisfactions.

Mumford addressed less decisively but some years before Tate the distinctions between unloving ubiquity and loving quality in communication (*TC*, 240–41, 274). With some debt to Victor Branford, he recognized the problems of an intensifying "chaos of stimuli" decades before Alvin Toffler wrote *Future Shock* (*TC*, 273, 284, 357, 433). He notes suggestively though without elaboration that the "human spirit," existentially cramped in the architecture, pragmatic routines, marketplace, and factory of the European early nineteenth century, rose "to a new supremacy in the concert hall" (*TC*, 203). Better yet, he imaginatively leads his reader into momentary phenomenologies of mining (contrasting mining, briefly, to farming; *TC*, 66–67), of the sudden availability of glass for windows, high-quality spectacles, and mirrors (*TC*, 125–29), of machinist artisanship (*TC*, 210, 324–25), and of the indirect emotional reaction to mechanization (*TC*, 284). He observes, ahead of most American education, that technics, in the hands-on sense, is an enormously valuable "instrument of discipline and education" far beyond parochial economic practicalities (*TC*, 321) although never a shortcut or substitute "for organic experience" (*TC*, 343). Technics as abstract study cannot replace organic experience because, as he puts it in the words of Waldo Frank about the similar case of art, it "cannot become a language, hence an experience, unless it is practiced."

Technics and Civilization breaks through deadly serious but relatively parochial American issues of the thirties to define the profounder problems with stunning independence and appropriately historical and cultural range. Mumford's own metaphors from the heart stand in provocative implicit dialogue with his ideological (quasi-patriarchal Geddesian? Branfordian?) lingo of taxonomic planning: "web of life . . . complex interplay . . . harmonious adjustment . . . ecological balance" (*TC*,

256). If he partly exemplifies the very problems he defines, he also identifies a prophetic agenda for his own work and that of the most ambitious cultural critics: "As our thinking becomes synthetic and related, instead of abstract and pragmatic, as we turn to the cultivation of the whole personality instead of centering upon the power elements alone, [we must deal with] the problem of tempo: the problem of equilibrium: the problem of organic balance: in back of them all the problem of human satisfaction and cultural achievement" (*TC*, 433).

The Culture of Cities was for 1938 an informed and informative book. Read after *Technics and Civilization,* it is also a somewhat groping, uneven, almost broken-backed production. It remains a provocative trial stage on the way to placing "the city in history," as he did in his masterly book of that title in 1961. The baldness of my list and sketch of the faults of *The Culture of Cities* will be alleviated by subsequent discourse—as it ought to be with regard to a book with such considerable riches. In the book there is a major shift of direction or mode. After an introduction and some two hundred pages or three chapters of history, on the medieval town, on "court, parade, and capital," and on the "insensate industrial town," the reader enters a substantial fourth chapter on the "rise and fall of megalopolis." Somewhere in the seventy-five pages of the fourth chapter, the reader is likely to feel no longer on a historical itinerary through urban streets but to be airborne on an arbitrary jaunt, with here a soar into a revision of the cyclicism of Patrick Geddes or into conceptualizations of region and there a touchdown in Amsterdam, in the Tennessee Valley, on Westchester Parkway, or in Welwyn garden city. In the flights and landings, what even in this age of text one may call voice changes more erratically and less coherently than it does in connection with the dialectic ambivalence of *Technics and Civilization.*[13]

13. "In going through *The Culture of Cities,* I find little I should like to revise in the first five [historical] chapters [counting the Introduction as one?]: with the last three, on the other hand, I have many kinds of dissatisfaction: they lack focus and they lack bite: even the part on the Garden City is not as good as my introduction to your reprint [of Ebenezer Howard's *Garden Cities of Tomorrow*]" (December 11, 1946; *Letters of Lewis*

Haste and fatigue may have been a factor, "a smudgy paragraph here, an overlabored thought there, or a too facile technical vocabulary in another passage," he wrote to Van Wyck Brooks (*VWB*, January 22, 1952; also in *W&D*, 269). He might have added, "occasional sentences featuring redundancy of the now/today sort, or syntax such as to smuggle noun-verb disagreements past copy-editing and proofreading eyes" (*e.g.*, CC, 279, 274, 492).

But a more fundamental difficulty for his writing about city, history, and culture seems to grow out of his notions concerning the relations between history and culture. Several such notions evidently competed in his mind with force verging on confusion—even if fruitful confusion. After a relatively short fifth chapter, predominantly general, on the "regional framework of civilization" and a sixth, sometimes more pragmatic and particular, on the "politics of regional development," he— deep in his long final chapter on the "social basis of the new urban order"—provides a remarkable gloss on the stressed notions of the earlier chapters in *The Culture of Cities*, and of chapters in other volumes of *The Renewal of Life*. In a subsection on the "mission of the museum," he praises museums as instrumentalities for "coping with the past" rather than (in words not his) dysfunctionally ignoring the past or dysfunctionally living with it and risking engulfment. In ostensibly enlarging more particularly on the museum as an element of the urban landscape which properly frames the past, he offers a rationale that expands to enormously general significance:

> One does not need a medieval house to appreciate a picture by Roger van der Weyden or Breughel, because, even if we had the house, we could not see either the environment or the picture through medieval eyes. On the contrary: the more complete the detachment and the more effectively we can screen a symbol from what it meant to another generation, the more swift and final is our response.
>
> Not unfortunately it is in a sense by our misinterpretations of the past that the past lives again: true understanding would leave the past precisely where it originally was: it is by its "otherness" that the past enriches the present with hints, suggestions, meanings that had no existence in its own day. (CC, 446–47)

Mumford and Frederic J. Osborne: A Transatlantic Dialogue, 1938–1970 [New York, 1972]).

Granted we cannot become medieval citizens; but it does not follow that we would not gain from having the contextual house. Granted that something good may happen—our perception of an artifact may be valuable to us—in the relative absence of context; but it hardly follows that the most value may come from the least context or greatest detachment. Mumford wisely avoids suggesting that complete detachment is possible. In saying "the more complete the detachment," he seems to mean "the more *nearly* complete." In one obvious sense, completely detaching a "symbol from what it meant to another generation" could scarcely avoid making it meaningless to this generation and presumably others. Mumford's thought implies at once a privileged unconscious or subconscious that can respond in a "swift and final" way to an artifact and a burdensome past threatening to occlude or eclipse the cognition that can come of the response. What is the burdensomeness of just what past? Mumford's ideas would seem to be most plausible for nonliterary artifacts. Yet, as we shall see, he construes architecture as a text thick with history. Still, it is not absurd to think of the past as living, for worse and better, only "by our misrepresentations," although it is a weirdly reductive notion of true understanding that requires comprehending an artifact fully without situating it in an ideational context. But that may be just a vagary of Mumford's expression: by guarding the "otherness" of the past, we *can* defamiliarize the present and deliver ourselves from the little death of dull familiarity or obliviousness to alternatives. But, again, can we know that a meaning we find in an artifact from another generation "had no existence in its own day"? Rarely; rarely; yet again, his effort to legitimize presumably unhistorical meanings in artifacts opposes a deadeningly positivistic historicism. The eddies of conflicted Mumfordian thought do not easily subside.

Amid the pulls and resistances, the central genius of the book rests (in Mumford's prefatory words) "primarily on first-hand surveys," on a deep reading in the "state of building [the] legible script, the complicated processes and changes that are taking place within civilization itself" (CC, 402–403). Here and elsewhere he insists on the experiential inclusiveness of the built environment as a symbol system: it represents a "large part of man's daily surroundings" (space and time), the reflection and focusing of a "wide variety of social facts" (history), a crystallization that is visible and "subject to the test of constant use" (kinesis, aesthetics, and practicality), an externalization and, implicitly, a reinforcement of the perceptual-aspirational elements underlying experience, and an adjustment of "human needs and desires" to the "obdurate facts of site, materials, space, costs." " 'Living' in architecture means an adequate relation to life, . . . not flowered wallpapers . . . but space and sunlight and temperature conditions under which living plants could grow" (CC, 412).

In response to the urban challenge, Mumford contends against signs banefully divorced from referents, as occur in "paper planning," where diagrammatic signifiers may square with meanings in the minds of the planners but fail "to grasp firmly all the elements" that "genuine planners" are alive to (CC, 375–81). In his pages on regional planning, as later in his italicized assertion that "as the background becomes more standardized . . . so will the foreground become more individuated" (CC, 454), he verges on a structuralist system: inclusive, interactive, and self-regulating.[14]

But elsewhere he treats closures of signification as abuses of the sign and referent alike, in that they amount to interference patterns between one sign system or lexicon and another. He "expressly" (but not altogether consistently) rules out "false biological analogies between societies and organisms" (CC, 303) and the contamination of the order of thought that privileges *economic* in long-term inclusive senses with that which privileges *profitable* in short-term, narrowly fiscal senses (CC, 339, and *passim*). He laments the "economist's naive myth that the gigantic metropolis is what it is merely because of . . . economic benefits or . . . geographic situation" (CC, 233, and *passim*)—for Mumford, an egregious underreading or translation into an inadequate lexicon of the felt signifieds (in that sense, the realities) of megalopolitan congestion, hinterland inanition, urban institutionalized desires, and the like. As a paradigm of the abuses and distortions of sign relationships in the urban lexicon, he offers at one point an extended and, in a book celebrating city life, a powerfully curious cadenza on stone as stultifying and deathlike. He manifests a passionate "distrust of the monument" (CC, 434–40). Mumford had come of age during the decimations of World War I, and he was beginning in the mid-1930s to foresee something of the decimations to come. Perhaps there is in his reaction to the monumental a very American element of death denial. Presumably the Hebraic-nomadic biblical tradition that "here is no continuing city"

14. As put by Jean Piaget in his still helpful primer, *Structuralism*, trans. and ed. Chaninah Maschler (New York, 1970).

claimed his respectful attention even if it did not win him over. He was reading the volumes of Arnold Toynbee's enormous *Study of History* as they appeared, and he appropriated Toynbee's idea of the historic tendency toward *etherealization* of "physical structures to their absolute functional minimum" (*CC,* 441; one thinks of banks evolving from fortresses to glass boxes to computer screens and printers—that is, toward nearer congruence of signified place, signified concept, and referent account balance). At once owing something to Ruskin and being his admonitory self, he cautions: "The protective function of the city . . . has been overdone. For living creatures, the only real protection comes through growth and renewal and reproduction" (*CC,* 442–43). He means that the traditional protective function of the city is now misconceived. He wants something like an *intimacy* of urban signifiers, and of their signified meanings in the minds of citizens, and of their referents in their lives.

Thus he can give qualified approval to the symbolic language of medieval churches and baroque courts as it becomes decentered, "dismembered," and "diffused" more at large in secular and semidemocratic forms (*CC,* Chapter II, Sections 8–16). Accordingly, too, he condemns arbitrarily geometrical grid platting as a "bad dream" among the "collective hallucinations" of paleotechnic urban planning and development (*CC,* 184–90); he reads and condemns the structures of paleotechnic laissez-faire capitalism as an all too apt translation of the mystical doctrine of divine right, turning the industrialist into the king (*CC,* 154–55); two decades later in the *New Yorker* he will read Colonial Williamsburg not as a salutary museum but as a "dead beauty preserved in embalming fluid";[15] he metaphorizes the inequivalence of medieval walled cities and capitalism, the latter a "cuckoo bird" that "could lay her eggs" in others' nests and crowd out the original "offspring" (*CC,* 22). All these are cases of displaced or attenuated or violated intimacy between systems of urban signifiers, meanings, and lives. Indeed, *The Culture of Cities* has deservedly enjoyed at least currency in this country and apparent influence abroad for exactly these sorts of discriminations, whether or not they have been identified in the terms I am using. The book, not unlike the text of Mumford's career, both presents and explicates conflicting languages of power and of sociable interaction. Although love is undeniably powerful, whether as *eros,* as *bienfaisance,*

15. Lewis Mumford, *The Highway and the City* (New York, 1963), 220.

as *agape* or grace, do we not recognize that love and power are for him, as for many another, the great opposites?

Mumford's own language varies. His rejection of *false* biological analogies evidently left him much that he felt free to use. He speaks of the city as a personality with grains, streaks, and fibers (*CC*, 481). He thinks of it in terms of quasi-genetic dominants, recessives, latents, emergents, mutants (*CC*, 390, and *passim*). He employs the term *organic* (*passim*) in senses ranging from the most literally physiological (as concerning bodily organs), through a physiological that shades necessarily in human cases into the mental, to the allegorical. But he essays explicit and literal definition and attends early and late to the inevitable problems of language and representation in doing so. "During the last few centuries," he argues, persons in power and many others have treated the lived experience of civic interaction as if it were an abstraction—mistaking referents and signifieds for nearly empty signifiers—and reified or referentialized "pragmatic abstractions such as money, credit, political sovereignty, as if they were concrete realities that had an existence independent of human conventions" (*CC*, 7). He partially defines the city as "in fact the physical form of the highest and most complex types of associative life," and as "life in its higher social manifestations" (*CC*, 482, 492). He as prophet-rhetorician would realign *that* signification with the signifiers that constitute the urban scenery around us, and vice versa. He foregrounds analysis in *The Culture of Cities*, foregrounds demystification in *The Myth of the Machine*.

One reason that Mumford continues to merit scrutiny is that he emphasizes the complexity of urban culture. In the complexity of cultural influences acting and manifesting themselves in the city, he sees a "perpetual guarantee against man's inveterate tendency toward over-simplification . . . that mechanization and falsification of the living reality which Bergson has correctly interpreted as a constant vice of human thought" (*CC*, 322). Farther away from Bergson, we would acknowledge the tendency, but we can define the vice as a function of desire abusive of language and failed by language. Even farther away from Miltonic images of concord than from Bergson, we can perhaps appreciate Mumford's celebration of "vitalizing . . . dissonance" in city culture, as opposed to the defensive conformity of many a suburb and small town, the debile restrictiveness of the old primitive and new exurban tribal village (*CC*, 488). Surely Mumford both cites and enacts dissonance in his cause of reinforcing the "cult of life" against the "artful system" (CC, 11). In quasi-epic roll call, he notes together some men of *insights* and of *work*, because, in his conception, urban works are texts and texts are actions—men including Peter Kropotkin, Patrick Geddes, Frederick Law Olmsted, and Frank Lloyd Wright. He signifies his own clear awareness of the scope and bearing of his celebratory-adversarial text by noting a kind of generic lineage: "It is not the triumphs of urban living that awaken the prophetic wrath of a Jeremiah, a Savonarola, a Rousseau, or a Ruskin" (CC, 6).

Much later he will write, "Until well into the 1930s we could always see the bright side of the darkest cloud" (SfL, 129). But by July 24, 1936, he wrote to Van Wyck Brooks, "When we were young we could ask ourselves: what can we conquer? Now we can only ask: what can we save?" In *The Culture of Cities*, he intermittently finds the prophetic wrath incipient in *Technics and Civilization*, wrath that is profound, co-herent, and pervasive in *The Myth of the Machine*. His is a public lan-guage of cultural indictment—that the god being worshiped is false, that the truth is other than conventionally thought. Mumford speaks a language of outrage but one undebilitated because sustained by an alter-native vision and by ego strength. In the role of the prophet as de-mythologizer, he refers to the "two orders of thinking, the organic and the mechanical. . . . The first springs out of the total situation, the other simplifies the facts of life for the sake of an artful system of concepts, more dear to the mind than life itself" (CC, 133). He asserts that "so-cial values and financial values are in decisive conflict" (CC, 391). And in opposition to an emergent "service-state," he deplores the "obses-sive mythologies and life-defeating mechanisms of the power state" (CC, 366).

His prophetic address can cool to the judicious, as when—in con-trast to his fervor in designating the "cult of death" (CC, 11) and the "life-defeating"—he glosses Freud's concept of the death wish as his-torically conditioned by the collapse of the Austro-Hungarian empire and the demoralization of Vienna (CC, 267–71). Of course, it does not make light of the death wish to tie it to collapse and demoralization, if those conditions have become constants—in the sense of universally re-curring features—of world culture. Shortly beyond his indignation at the "cuckoo bird of capitalism" that lays its eggs in the nest of the medi-eval city, he notes that the baroque scheme of life "was lacking . . . suit-able provision for work" (CC, 134). That observation leads, though surely not inevitably, to another: "Just as capitalist enterprise itself was partly a protest against the stale privilege and staler routine of the guilds; so the cult of luxury was a protest against . . . miserliness and middle-class thrift" (CC, 135). The very syntax of this logical-historical balance of claims betokens a judiciousness existentially far from the impassioned end of Mumford's prophetic range: "B is to A as D is to C"— like Andrew Marvell.

Similarly, the truth was obscure enough to many to need pointing out when he wrote that the "Machine Age . . . has retained the steam-heater to produce sub-tropical heat at great expense, when it should have invented better forms of permanent insulation, to do away with extravagant heating of cold walls and windows" (CC, 442). But it was tendentious argument and near-petulance to phrase the failure as if it were decisive or even enduring, when actually it was the historic threshold of the asbestos disaster and the fiberglass triumph, not to mention of far more efficient modes of heating, all just a historical moment later.

Are we to make it out as crackpot petulance when, in the same place, he complains that the "Machine Age has treated machines as ornaments: it has invented the vacuum cleaner when it should have done away with the rug." A more extensive passage to the same point (TC, 277) indicates his sober seriousness. There is no frivolity in the rhetorical pointedness; there is more than petulance. The matter is instructive as to the inconstancy of Mumford's strength.

In Technics and Civilization, he asserts that the "appropriateness for use in interiors" of the carpet or rug, "if it did not disappear with the caravans where it originated, certainly passed out of existence with rubber heels and steam-heated houses." Certainly? The stressed and vulnerable human foot is, it seems, much better adapted to grassland than to flat rock. Sound bounces harshly off ceramic tile or hardwood but not off carpet, and to a lesser degree so does light. A fine rug is a splendid artifact and is portable as a parquet floor is not. Mumford's empathy failed there, not uniquely, but not characteristically, either. He was on that point of the compass a little mad perhaps; at least he was not historical enough and not, if one may put it so, phenomenological enough, and it is not simply irritation with the hucksterism of vacuum cleaners and the like that occluded his wits.

The problem, I think, was a vulnerability to occasional seductions by the narcotic power and pseudodecisiveness of a reductive or positivistic rationalism. Wonderfully immediate and experiential with regard to greatly diverse urban and rural phenomena—see the celebratory cadenza on the late-medieval festival procession (*CC*, 62–64)—he occasionally flagged. Mumford the American then lusted after the magic-utopian solution.

At times what he did seems partly an intellectual genuflection to his mentor Geddes—after whom the Mumfords named their son—in preference to Oswald Spengler or Toynbee. He promulgated, with revisions, the disturbingly arbitrary, distorted, and biologistic "cycle" of urban-cultural development and dissolution: eopolis, polis, metropolis, megalopolis, tyrannopolis, nekropolis (*CC*, 285–91). But almost immediately we find him hedging: these are "*logical* stages," not the "living reality." And the "life course of cities is essentially different from that of most higher organisms" (*CC*, 292, 294). Thence he moves on to the express ruling-out of "false biological analogies" (*CC*, 303). *Logical* is perhaps less apt a word in his context than *phantasmal*, but such interpretation would come later.

When he writes of, for example, a "consensus . . . gradually being established among men of good will and effective competence" in a context of "positive knowledge" (*CC*, 415), when he writes of regions as "deliberate works of collective art" (*CC*, 348), but does not indicate who will appoint and who will jury the artists, when he writes of apportioning publicly owned land (meaning all of it!) according to the "stable needs of the community" on a fifty-year horizon (*CC*, 327–29), and when he writes of all the things we *must* do "from generation to generation by the desires of the many and the wisdom of the competent" (*CC*, 478), then I think one senses a little uneasily the prophet as world gamesman and, even less appealingly, as czar.

Perhaps he underestimated society's ability to develop ways and means without the czarlike control of planners—by naïve definition benign—who would counter capitalists and associated mechanists, benign planners whom his imagination tended to endow with czarlike power. Certainly he discerned, in the burgeoning European fascisms of the thirties, mass psychoses of projected or displaced fear and hatred, and he rightly saw that those could be conceived in other power societies and malignly nurtured in the medium of city culture. It is hardly too much to say that whatever is healthy in any society has since the thirties been resisting movements toward megalomanic and rancorous violence, which he called barbarism. The widely varying degrees of success and failure in the resistance, and in some cases the uncertainty about the result, attest to Mumford's relevance. Toronto, Washington, Tokyo, London, Buenos Aires, Paris, Moscow, Beijing, Santiago, Damascus, Teheran, and Phnom Penh exhibit radically different degrees of illustrative potential for his strictures against cults of aggressivity in "A Brief Outline of Hell" and "Phenomena of the End" (CC, 272–83) a half-century after he voiced them. But none of those capitals exhibits zero degree. Mumford appraised as the "*deepest need* of modern life . . . to re-integrate the organs of association by forming new civic wholes" (CC, 470). And in a persistent note of implication throughout *The Culture of Cities,* he presented that as a crucial, delicate, and difficult problem of scale and distance.

A provocative analogy of Mumford's suggests that a region is often indistinct in its geographical boundaries but relates to its own inside and outside as a person relates to self and world (CC, 369). Compare the similarly oriented insight of Tate that a true region is more a matter of coherent history than of coherent or rigidly demarcated space. Mumford asked for equivalents of the New England town meeting, the "human scale in government" that can foster the "processes of persuasion and rational agreement" (CC, 142, 382). With a nod toward the local or private end of the spectrum, this man who relished both human communion and the solitude necessary for writing added (a little uneasily?), "If friendship requires a degree of isolated communion, so does neighborliness" (CC, 251).

Nagging intimations of the difficulties in achieving the most humanly fruitful scopes and distances give *The Culture of Cities,* not to mention the whole of *The Renewal of Life,* some of the prickly animation that redeems the book's somewhat doctrinaire injunctions for state planning and the state's ownership of land. The awareness of such difficulties frames Mumford's applause for an Americanism of personal individuality coupled with performance—of the "virile personality . . . celebrated by Emerson, Thoreau, Hawthorne, and Whitman . . socially equal . . . differences . . . no longer merely historic congealments but psychological facts" (CC, 448–49). His discountenancing of the personality's reduction to historic congealments, which occurs quite near the end of the book, underlines one of his ambivalences, and as such one of his animating themes: not so much the bland, Where is a usable past?

that Van Wyck Brooks might have had him ask but something more
like, How shall the past be not a dead but a helping hand, not a crypt
but a seed bed? [16]

"Modern man in the West first took shape in a period of cultural disintegration," Mumford begins the first chapter of *The Condition of Man*,
after a surveillant introduction, and he follows with ninety pages to
bring Western man up to the year 1000 and to carry his analytic exposition to the fourth of his eleven chapters, entitled "Medieval Synthesis."
That gives way, according to his argument and chapter titles, to "Capitalism, Absolutism, Protestantism," to "Uprising of the Libido," to "The
New Hemispheres" (on new spaces in several senses),[17] to "The Insurgence of Romanticism," and to "The Progress of Prometheus."

One can quickly recognize in the 235 pages of Chapters Four
through Nine a relationship with the first three chapters, almost as long
in total, of *The Culture of Cities:* "Protection and the Medieval Town,"
"Court, Parade, and Capital," and "The Insensate Industrial Town"—
and with the first two hundred pages, through eotechnic and paleotechnic phases, in *Technics and Civilization*. The earlier books focus
on how, to paraphrase Winston Churchill, we first build our cities and
our technics and they then build us. The third book in *The Renewal of
Life*, actually Mumford's eleventh book, addresses causes less material
and often less immediately efficient than those the earlier books do. Civilization for Mumford is culture materialized, and here he is concerned
with the *roots* and *dialectic* that issue in culture. Both metaphors are
fair to Mumford, and I should add that my invocation of Aristotelian
causalities to make a distinction should not obscure Mumford's stead-
fast opposition to determinisms.

But capabilities are not necessarily realized. The condition of Western man, whose past has potent connections back to Plato and Isaiah,

16. Brooks characterized the book as *hopeful, powerful,* and *healthy,* in a letter of
April 12, 1938 (*VWB*).

17. These spaces are beyond "walls" of class, occupation, fixed duties and obligations, and cities and territories. "Space summoned them to movement and . . . devoured
space" (*CM*, 231). It is somewhat pneumatic, aerophagic history.

represents for Mumford a cultural devolution. The worship and quasi worship of successive central "idols" has left us, despite initial satisfactions in each instance, like the man dispossessed of a demon in the biblical parable. Mumford does not draw the analogy but does define us as dispossessed of the demon of cynically demoralized classical culture, only to be entered subsequently by several other demons—"and the last state of that man was worse than the first." An idolum is an imaginative and ideological construction, typically less than a world view, but a more or less explanatory frame enabling action (*CM*, 231, and *passim*)—what later writers, and increasingly Mumford himself, would call a myth.

Because the book does wander in a kind of sociological twilight of idols, looking for causes of man's condition while remaining ambivalent about causality—and because there are other divisions in mind, which will come to notice—it is an instructively *limited* success. For its subject matter, one would prefer to read José Ortega y Gasset or Eric Voegelin. But one might approach this book and *The Conduct of Life*, as articulations of war-world American malaise, as instructive generic misfires, as intermittent recognition scenes.

So, to put it another way, this book, implying the earlier two, is an analytic account of the coming to present circumstances of *animal symbolicum Homo faber*. Of *symbolicum*, more in a moment; of *faber*, it must be emphasized that Mumford's belief in the significance and value of work empowers much of the best in the book. He virtually begins the volume with a summary of and overture to that belief, which was a theme of the two previous books in the series:

> Man gains, through work, the insight into nature he needs to transmute work into artifacts and symbols that have a use beyond ensuring his immediate animal survival. The ultimate justification of work lies not alone in the performance and the product but in the realm of the arts and sciences. The *role* of work is to make man a master of the conditions of life: hence its constant discipline is essential to his grasp of the real world. The *function* of work is to provide man with a living: not for the purpose of enlarging his capacities to consume but of liberating his capacities to create. The social meaning of work derives from the acts of creation it makes possible. (*CM*, 5)

That is almost to say, as Milton's Bard did of unfallen Adam and Eve, "Their work declares their dignity." Mumford's eye, in conceiving his argument, fastens alertly on corollaries of that complex and central conception. One: that although science can be variously a curse and a blessing, in the laboratory "man the worker became companion to man the knower" (*CM*, 248). Another: that the labor of "book-learning [is] a veritable Baconian 'acceleration' of mental germination" (*CM*, 257). I shall return to the metaphor of growth.

Certainly, too, Mumford recognizes human work as *respite*, and potentially as neurotic displacement or avoidance, as *decadent* (in the definition by Johan Huizinga, whose writings Mumford applauds), not because of what was there but rather because of what was left out:

> Once the [early monastic] novice made the final surrender, he was free from the burden of intolerable accident. . . . One day was like another: thanks be to God! . . . The uniformity of monastic existence touched every detail. Monks were the first wearers of uniform costume in modern times: by comparison Michelangelo's uniform for the papal guards was a parvenu. More than that: regularity and order bore fruit in economic practices. In every department of work, order, repetition, standardization are great economizers of labor. These monastic habits, spreading into every part of estate management, brought their natural reward in increased wealth. Not merely is Coulton correct in describing the Benedictines as essentially the founders of capitalism, but it is equally correct to add that the business man, the bureaucrat, and the mechanical worker are all the specialized end-products of monasticism. Some of the life-denying practices of modern capitalism had their origin in the withdrawal and retreat of this early epoch. (*CM*, 94)

Indeed, in the first subsection of "The Progress of Prometheus," Mumford defines the "utilitarian ideology" as offering "an idolum *and* a social pattern that responded wholly to [extroverted] personal needs . . . the very possibility of throwing one's whole life into the systematic exploitation of nature . . . the counting-house and the factory" (*CM*, 301). Characteristically, his point here has to do with what is left out or occluded (for example, parks and gardens; Mumford's attitude toward wilderness was rather indifference than Sierran affection or Eiseleyan ambivalence).

Similarly, he writes, "The economic motive was lacking. . . . Hence the most important problem of all was left out of the new world-picture of science: who is to control the controller of nature? . . . The civilization so created, we have now learned . . . utilizes its full energies only in war" (*CM*, 244). More generally: "Mechanical time was non-cumulative: it accompanied only changes of motion, changes of position," excluding "history, as the cumulative enregistration of experience" (*CM*,

247). More profoundly, he assails the illusion of "immediate man" as it is implicit, for outstanding example, in the skepticism of David Hume: "The real Reign of Terror began: the beginnings of a nihilism that has reached its full development only in our own times. . . . Hume's essential doctrine was the autonomy of raw human impulse and the absolutism of raw sensation . . . a bald sequence of abstract sensations in time" (CM, 269). Mumford sees, though he can momentarily forget, that there *can be no* unmediated man.

Not perhaps similar at first sight but a crucial counterpart to the stricture against science for occluding framing considerations is his ending to a subsection on Marxian thought (CM, 329–42). He closes a reflective appraisal of Marxism's discernments, contradictions, and denials with a suggestive historical-psychological formulation that later contributes some of the force to *The Myth of the Machine:* "Socialism had been framed as a working-class counterpart of the utilitarian ideology; for Marx's mission was not merely to turn Hegel but also Ricardo upside down" (CM, 341–42). Economic activity and symbolic activity are not always, and are never merely, sequential or dialectical. They are cotemporal, and either may be a *basis* for the other, without necessarily being a *cause*. It is not clear (at least to me) that any progress made since Mumford in further untangling this knot has not been nullified by widespread regress.

As Mumford is almost always the splendid neighbor and fellow citizen with regard to a city he has lived in or a piece of equipment he has used, so he is also with any program of thought aptly represented by someone's writings. Generous spirited but with sharp eyes for fallacious thinking and dangerous tendencies and, as is plain, for what is left out, he is at least suggestive, and can be provocative and valuable. His Hegel was perhaps an American Hegel of the early 1940s, that is, sketchy, and his Augustine more fierce and antiworldly than the diverse-minded rhetorician of faith, hope, and love that he seems to me, but no matter: Plato, Ignatius, Calvin, Marx, Darwin, and Freud do not make cameo appearances but have, so to speak, regular guest engagements—and Mumford's mind engages them provocatively. One outstanding example is Freud; Mumford insists throughout the book that Freud was dysfunctionally wrong not to recognize a positive role for the superego as well as a negative one.

An ambitious and admirable effort it is to be teacher in residence to citizens of influence. Gallant, it is, too, given the odds against success when the general argument to fellow citizens must run that "the education to which you have committed yourself is too narrow; the values to which you have habituated yourself are deathly." Still, Mumford's control can falter when he is out of a city or a body of writing in which he may be said to have *lived* and is obliged to generalize on broader and more amorphous ground. Signal instances of that are evident in his reasoning about mentation and expression, about the historical fact of Christianity, and about the complex of death-time-history, as well as in the *locus desperatus* of his own omnipresent term *organic*. Each of these areas invites brief separate consideration. To give them that with some care for Mumford's sequence of presentation is almost, in the manner of *Rashomon*, to add four alternative narratives to my initial sketch of his argument that culture has been devolutionary. But it is not inappropriate for that.

Late in the next to last chapter, "Barbarism and Dissolution," Mumford avers that "a dynamic syncretism of doctrine and creeds and philosophies was needed . . . if the idolum of the machine was to unite under a new sign with the idolum of organism" (*CM*, 381). That hot-air balloon of "dynamic syncretism" is ostensibly pegged to solid ground by the example of the social activist and biologist Patrick Geddes. But not firmly enough. More promising is his implicit concern with education; that is a recurrent motif in this and others of Mumford's books, as it was in his life, which included the nitty-gritty of school-board service. At the conclusion of *The Culture of Cities*, he half-proposes, half-identifies the "school as community nucleus," the architectural embodiment of "purposive group life" succeeding such earlier embodiments as—in historical sequence—church, court, factory, and market (pp. 471–75). The new city, he adds, can become the schools' chief instrument; this was thirty years before the deservedly esteemed Peter Drucker promulgated the doctrine of the knowledge society.[18]

More promising for Mumford's subsequent writing of *The Myth of the Machine*, and indeed more rewarding in his analytic retrospection in *The Condition of Man*, is what he has to say on the matter of idola and the unconscious or semiconscious needs and desires they serve. One of the final paragraphs of "The Basis of Renewal," and thus of the book, begins, "Today our best plans miscarry because they are in the hands of

18. See especially Peter Drucker, *The Age of Discontinuity: Guidelines to Our Changing Society* (New York, 1969).

people who have undergone no inner growth" (*CM*, 422). He elaborates: "Their hidden prejudices, their glib hopes, their archaic desires and automatisms—usually couched in the language of assertive modernity—recall those of the Greeks in the fourth century B.C. or those of the Romans in the fourth century A.D."

Mumford wrote in the generation that saw the German publication of Ernst Cassirer's *Philosophy of Symbolic Forms* and other works. *The Condition of Man*, in 1944, annotates Suzanne Langer's *Philosophy in a New Key* (calling her Langner) as "admirable exposition of symbolism and semantics" and recommends it. Cassirer's name does not appear in *The Condition of Man*, but his *Essay on Man* is in the bibliography to *The Conduct of Life* in 1951. Mumford includes in that bibliography Ortega's *Toward a Philosophy of History*, with the pregnant comment that "the view of human life as a fabrication and a drama parallels at more than one point the underlying theme of the present series." [19]

Whatever the interaction of the directive influence, provocation, or reinforcement from his precursors and peers, and his own independent insight, he wrote with a sense of the range in mentation. His *Homo faber* as *animal symbolicum* takes some position between the extremes of dream and of "positive knowledge." (Comte is "magistral"; Santayana's "late masterpiece of classic humanism," *The Life of Reason*, is "almost a museum piece.") He not improperly cites a Reformation figure:

> The failure to understand the role of symbolization in human life has been responsible for a grave misunderstanding of the nature of man. Because symbols are subjective in origin, in that they are not found in nature outside man, many people fancy that they are unreal, mischievous, or that a more sound existence would be possible if all symbols were excluded except those that could be reduced to quantities or visible operations. Those who have advocated this view lack an understanding of man's essence and true aptitude. On this subject Jean Calvin, for example, is a safer guide than many current behaviorists. "The manifold agility of the soul, which enables it to

19. Suzanne Langer, *Philosophy in a New Key* (Cambridge, Mass., 1942); Ernst Cassirer, *Essay on Man* (New Haven, 1944); José Ortega y Gasset, *Toward a Philosophy of History* (New York, 1941). Other works with some relevance here: Carl Jung, *Modern Man in Search of a Soul* (New York, 1936); Sigmund Freud, *Interpretation of Dreams* (New York, 1955); Alfred North Whitehead, *Science and the Modern World* (New York, 1925); and Erick Kahler, *Man the Measure* (New York, 1943).

take a survey of heaven and earth; to join the past and the present; to retain the memory of things heard long ago; to conceive of whatever it chooses by the help of imagination: its ingenuity, also, in the invention of such admirable arts, are certain proofs of the divinity of man." Thus Calvin. This agility of the soul is the result of man's development of the symbol: a more miraculous tool than the fire that Prometheus stole from heaven. (CM, 9)

Certainly he recognizes and insists with some ambivalence that the agile soul conceives of things chosen and unchosen.

Mystery and darkness envelop man, and night to him has many forms, the night of his animal past, the night of his unconscious urges, the night of an ignorance whose circumference seems to widen with every expansion of his conscious knowledge, and finally, the night of non-existence and death, which encloses his personal life at both ends. Reason could gently laugh the Homeric gods out of existence, because they were inferior in their sudden angers and exorbitant lecheries to the self-control practiced by living men: but reason could not exorcise the demons that rose out of man's unconscious, nor could it extricate man from a chain of events in which both determinism and chance sometimes mocked the efforts of the human will. (CM, 24)

He was, after the coming of war in 1939, increasingly both a herald of widening darkness and a herald within that darkness of something quite other. Historically, "there is scarcely a single aspect of [Romanesque] culture which does not become clearer when one interprets it as a neurotic dream phenomenon" (CM, 107). Again, in medieval Europe "dreams governed" (with reference to hellfire, infant damnation, and bodily resurrection; CM, 139). Or, in the wake of Comenius, "the underlying aim of New World education will be the fabrication of Mechanical Man . . . who will accept the mechanical world picture . . . will submit himself to mechanical discipline . . . a dream no less than the old Christian dream of angels, sprites, demons" (CM, 259).

Obviously Mumford uses *dream* in an expansive sense, from the familiar sleep phenomenon to the *somewhat* more consciously orchestrated waking mode indicated by the subhead "The Mechanical Idolum." Such usage is not unconventional but can be tricky. A trickiness he insisted upon against Freud's "uncritical mediocrity as a philosopher" is the validly discerning and revelatory possibilities of dream, and the positivity of the superego (CM, 363–65). That dreams, impulses, and intuitions welling up have helped man give part of the outer world the dimension of his own being is a partly Kantian recovery that positivism unreasonably condemns, Mumford virtually says.

A form of trickiness in symbolization which he fitfully recognizes, a trial of the sign from which he might now and then suppose to recuse himself, is the historicity of representation. Against the world of "primary qualities" common to rationalist and empiricist, he rejoins:

> But man is not born into that bare physical universe: rather, he is born into a world of human values, human purposes, human instruments, human designs; and all that he knows or believes about the physical world is the result of his own personal and social development. The very language he uses for neutral scientific description is a social product that antedates his science. Indeed, the tendency to look upon processes in the physical world as more important, more fundamental, than the processes of organisms, societies, and personalities is itself a by-product of a particular moment of human history: the outcome of a systematic self-deflation. (CM, 10)

Similarly, he praises Luther for enunciating the "essential truth of religion, namely that it is a relation between the self and the divine purpose . . . a relation forever endangered by the visible symbols and allegorical references man creates in order to further his own understanding of this mystery" (CM, 185).

Mumford's problem is the problem of anyone trying to present a great and dynamic thing in signs. Not only will the given signifiers surely provoke somewhat different signifieds in the minds of diverse readers, not only will even the corporate share of that sign bear a problematical relationship to much of its referent in history, but there is the problem of simultaneity—the more difficult and inevitable, the more complex and dynamic the matter. Witness Milton's Raphael trying to tell Adam and Eve of simultaneous divine justice and mercy, which to human ears "cannot without process of speech be told." Witness Mumford's commendation of the "essential truth of religion," which for his situation as (shall we say?) Koheleth, for the significance of his preachment, and our well-being as readers needs to be understood as both framing and being framed by, among other passages, one thirty-eight pages earlier: "Our natural order is but Dante's world in reverse; and our task is to bring a sense of the eternal and the infinite back into our common daily life . . . not by claiming false knowledge about the unknowable, but by understanding that our life, too, is an allegory, because what man makes of it in the soul is ultimately as important as what is given in the senses" (CM, 147). So the effort of symbolizing coherently must always by implication go on, never, after Babel, quite catching heavenly presence in representation. It presents perennial occasions for human creativity and, in the retelling, Mumford's generosity. His prophecy, like any true prophecy, is hermeneutic: beginning in the middle with questions, looking back because the end cannot be understood save in the light of the beginning, and looking ahead because the past and present cannot be understood save in the light of the end.

Two instances: "The very development of naturalism would have been disordered and unprofitable without the abstract sense of order that theology itself had created" (CM, 143); "Capitalism was often in its early phases a healthy, liberating influence" (CM, 184). And with regard to civic ensembles and other things, he is disposed to honor what Coleridge called the coadunative power of the symbolic imagination. Yet in at least one instance he seems to hanker after a curious presymbolic or extrasymbolic secular "magic":

> The monks elaborated if they did not invent the mass. By a steady refinement, they created a unified whole out of the procession, the choral chant, the vocal prayers and responses of the entire congregation: all enhanced by the smell of incense, the burning of candles and lamps, the solemn familiar words, the dark high hall. . . . Even in a tawdry church today, cut off for a moment from a dusty street loud with the screech of brakes, the whine of motors, within an interior set with hideous sculptures and painting turned out by the gross lot, even here the mass has a power of evocation that goes far beyond the beginnings of Christianity: it carries overtones of an *unintelligible ritual magic* that perhaps existed as early as the dawn of human speech. Long before the modern painters had explored the values of abstract art, the Church had created an abstract and depersonalized art which penetrated the *recesses* of human feeling far more powerfully than words and gestures better understood. (CM, 95; my emphasis)

"Perhaps existed as early as . . . speech." Why not before, if "unintelligible"? Is Mumford's remark a guarded act of faith in something fundamentally human yet outside the human symbolizing activity of which he insists language and the city are the richest expressions? If so, perhaps he is right, but such a surprising—albeit secular—profession of faith scarcely admits verification or falsification. Yet again, it suggests a

Mumford hermeneutically resistant to premature closure, resistant indeed to quasi-totalizing formulations.

To give *magic* so apparently honorific a sense is uncharacteristic of Mumford, though he will give a carefully qualified approval in his own ambiguous imagery for the primordial. He applauds naturalism, in one major instance, for countenancing, against the "shallow rationalism of the mechanists," the "*residual* impulses, *welling* out of a *blinder*, starker past, to which theology had affixed the name of 'original sin'" (*CM*, 351). Certainly the matter of "original sin" is vexed, and Mumford's rejection of that trope of symbolic discourse need not occupy us.[20] But one does wonder if there is not an irrational component in the counterposing of darkness, odoriferous recesses, residues, and blind welling—fecund? fecal? who knows how?—against the scientific "positive knowledge" that "replaced mere opinion, as rational opinion had replaced authority" (*CM*, 377). Such "positive knowledge," whether imaged geologically or politically, seems to enjoy hill-country reference-point rigidity, and better drainage. In one especially bland and blithe passage about science never taking a step back, he avers, "Even fundamental over-all changes in science . . . still leave the *constituent increments* of verified truth intact: they serve like permanent *officials* who carry on no matter how radically the government *itself* may change. . . . Science was essentially an open synthesis, open at both ends, at the top for each *accretion* of *fresh* knowledge, and at the *bottom* for each necessary modification of its own metaphysical and logical *substratum*" (*CM*, 251–52; my emphasis). Fresh rock or plate tectonics, nothing fetid, here. Correspondingly, bright air and water beckons:

> Clarity at the expense of life. But what clarity! Read Galileo, Descartes, Spinoza, Newton, Locke: it is like taking a bath in crystal-clear water. Their universe is clean, neat, orderly, without smells, without flavors, without the rank odors of growth, impregnation, or decomposition: above all, without the complications of real life. Their maps show the bare physical contours of the landscape: if one remains in the air they are adequate. But this new world picture gives no hint of the soil, the bacteria, the mat of vegetation, the animal life: it retreats from the dense atmosphere of actual experience to the stratosphere of its own rarefied abstractions. All the forms and processes of reality, to which other ages had given a full, if muddied, expression by means

20. See the brilliant essay on original sin by Paul Ricoeur in *The Conflict of Interpretations: Essays in Hermeneutics* (Evanston, Ill, 1974).

of fable, superstition, myth, allegory, now dropped off in the fresh water bath of science. (*CM*, 260)

The ambivalence about the lively (tending toward the fetid and feculent) will reclaim our attention when we turn to that most magic word in Mumford's own secular liturgy, *organic*. The—shall we call it *intermittent?*—affinity to the clear (hence visual), taxonomic (hence tending toward the lifeless) will reclaim our attention when we turn to his concern with death, time, and history.

It may be noted, as entries in this sketch map of Mumfordian cultural geography and of stations in his liturgical sociology, that he more consciously engages certain metaphors situated comfortably between the ambivalent extremes of clear taxonomy and dark welter. One, throughout, is of world as theater, as in "*The scene changes*. This device of the baroque theater now had its counterpart in life [in] exploring and pioneering and colonization" (*CM*, 238). He characteristically foregrounds the elements of dynamism, symbolization, and choice implicit in the metaphor, occludes the elements of pretense or deceit, on the one hand, and of determinism on the other, as he does also in a judicious cadenza on Romanticist utopia as *fête champêtre* (*CM*, 296–300). His own utopianism, and book title, stem however exfoliantly from Patrick Geddes: "'Animals are always attempting the impossible and achieving it.' That, he would add, is the essential condition of man" (*CM*, 389).

The sociological metaphoric use of the technical biological terminology of *dominant, recessive, mutant,* and *emergent,* and of parasitism seems to represent a none too conclusive attempt to characterize the dynamism of social life with a decisiveness borrowed from biological "positive knowledge." The same is the case, though more vaguely, with *balance*. It is notorious that one man's system is another man's subsystem, oriented or even rendered contingent by the different frame. Indeed, it may be that one man's system is another man's Other. The Mumford of *The Renewal of Life* seems to believe, perhaps with decreasing assurance—rather as the brash young Donne had put it in the 1590s—that atop a "huge hill, cragg'd and steep, Truth stands" and may be won, albeit with much going-about. Only later would Mumford approach the more guarded position of Milton in 1644 or of that belated seventeenth-century man Loren Eiseley—the former that truth is like the fragmented body of Osiris, never fully reintegrated until judgment day, the latter that human systems or bodies of knowledge are like hummocks of quasi ground in a bottomless and boundless marsh.

Of course, Mumford has particular difficulty with Christianity, a historically manifold system that in certain respects resists categorization as a system at all: making an absolute claim on faith, it is conjoined with the conviction that anyone's faith will be—like everything else in a *fallen* world—imperfect and that, faith in a deity by definition exceeding definition and accordingly transcending any human system, including supposed polarities like life and death, it will certainly transcend any theology or any church. That formulation is not Mumford's, but he acknowledges Christianity's unique and even mysterious consequentiality for the West (*CM*, 52–75).

The evident problem of Christianity for him is not merely the familiar—and one might say Christian—distinction between the man of Nazareth and his teachings as institutionalized in what Mumford all too reifically writes of as "the Christian Church." Of course "the Church failed" (*CM*, 56) the man of Nazareth; *they*—the churches—have always failed him, as much ecclesiological theology has tended to insist; and incessant arguments and divisions within the church from the very first have reflected both conscious and unconscious engagement with that fact. Many theologians would accept the characterization of the church as a "gigantic tomb" (*CM*, 63) in the absence of divine grace, though some might prefer Jesus' phrase *whited sepulcher*. A Christian theologian would believe that divine grace has from time to time operated, and indeed if at all liturgically minded would believe it to be individually available, though in no magical sense, in the Eucharist. Milton, no priest, believed in "grace prevenient" without which no one could even pray; and so on.

Mumford's own tenets of faith are to the contrary with regard to "the Church," the "gigantic tomb," and almost equally to the contrary with regard to a similarly reified "Christian":

The Christian is a person who rejects the usages of a dying society, and finds a new life for himself in the Church. He overcomes the local forces of dissociation and disintegration by attaching himself to a universal society. He builds his life around the themes of rejection and succorance; and he balances all his temporal difficulties against the hope of a divine justice that will punish his oppressors and give him a share in eternal glories. When Christi-

anity came to be defined in these terms, it should have been apparent that
Jesus of Nazareth was the first heretic. (*CM*, 75)

"Dying society" was not so monolithic either, but that point may be left
to the historians. "New life . . . universal society," *perhaps* "rejection
and succorance"—well enough. But Mumford's *balances, all,* and *punish* tendentiously create a straw Christian. Just possibly his name was
legion, but from Saint Paul's times onward there have always been numerous Christians to recognize that in those terms Jesus was the first
heretic. Mumford's voice is not nearly so lonely as he apparently needed
to believe.

More extremely, more idiosyncratically, he avers a few pages later
that "seeking holiness, above all seeking peace, the Christian finally
built a self-contained life around the themes of rejection and death"
(*CM*, 89). Which Christian? The context is the early appearance of the
contemptu mundi tradition, the desert saints, and in the following paragraph, Gregory and Jerome as they appear in supportively extremist
cullings. Syntax could in that way wrench diction to align *Christian* and
Simon Stylites. But I suggest that this is also the *Christian* of psychological convenience for the ambivalent Mumford. He wrote in July 12,
1942, to Van Wyck Brooks, of "the corruption of liberalism . . . the
general disease from which our civilization was dying. If I had suspected
how deep it went, I would have given up all thought of pursuing my
literary career in 1937 and turned seriously to politics: because it is for
lack of political fortitude that we are now in such a desperate condition."[21] But a writing desk in Amenia, New York, is not too fugitive and
cloistered a virtue compared with an Egyptian reclusion of sand, offal,
and empty sky. Modern, and American—though far from exclusively
so—is Mumford: "I should have been farther up front, closer to the
shooting."

His own ambivalence about the ambivalent Judeo-Christian concept
that the things of this world both do and do not matter is quite palpable.
It ranges from the shakily contentious ("the crucified God, a being born
of the worshipper's own sick needs and ambivalent desires"; *CM*, 57)[22]

---

21. Compare his earlier remarks to Brooks: "A half dozen of us, firing steadily from behind our book barricades, may yet keep off the fascist hordes!" (*VWB*, July 24, 1938). "In a
time like this [Hitler-Stalin pact], salvage is almost creation" (*VWB*, November 3, 1939).
22. Wolfhart Pannenberg notes that "God is not anthropocentric if the very idea of

to what might be called the ironically revealing: "The vital lies of the Christian's creed were a better answer to the desperate and the disheartened than the numb truths of the pagan philosophers" (*CM*, 69). And, "Those who are contemptuous of Plato's uncompromising 'idealism' might ask themselves how many power states have lasted as long as the Benedictine order" (*CM*, 93). Is it necessary to observe that the Benedictine rule represented, properly, a *measure* of withdrawal conjoined with hard work, all ordered in the name of love (perhaps seeming an actual or potential analogue to, say, a household in Amenia)? "Vital lies . . . numb truths" betokens a faltering sense both of how faith statements, banished from the lexicon of "positive knowledge," marvelously reappear in the inkwell of diction and of Mumford's own language's historicity, and the creed's.[23]

It is estimable of him to praise the psychological acuity of Ignatius' *Spiritual Exercises* a decade before Louis Martz's *The Poetry of Meditation* widened awareness of their literary and imaginative significance, and it is understandable that he finds, on the other hand, that "the Society of Jesus was conceived under the sign of the Despot" (*CM*, 230). But it is curious that the Mumfordian mind is driven beyond reification or questionable argumentation or psychological projection to a sort of compulsive reductivism:

> Francis was in love with life and he lived a life of love: that example was beautiful in itself, but ineffectual for the community. What Francis actually achieved as an organizer of Christian life could have been accomplished by a Dominic: did not Dominic in fact accomplish it? . . . Even that love which prompted Francis to wash the sores of lepers was less effective, indeed less kind to afflicted humanity, than the new medieval medical practice of isolation, which, though it might narrow the province of love, also lessened the incidence of this disease and finally wiped it out. The love that Francis preached needed the support of positive knowledge, economic foresight, rational order, so that it might also take firm hold of the university, the

personality is divinely inspired" ("The Question of God," *Interpretation*, XXI [1967], 302). And René Girard has pointed out that the Jews seem to have been unique in engaging *sympathy* for the sacrificial victim (*Things Hidden Since the Foundation of the World* [Stanford, Calif., 1987], 144–58; *The Scapegoat* [Baltimore, 1986], 100–111).

23. For a placement of private judgment in ostensibly equitable counterpoise with public judgment, but grotesquely skewed by a foregone conclusion (supposedly a case in point) from plane geometry, see *CM*, 195.

guild, the court of justice, the parliament. As an isolated and dissociated love, Francis's divine love, like Tristan's carnal love, was doomed to defeat. (*CM*, 126)

Ineffectual, accomplished, effective—these seem to signify convenient administrative and public-health countables (important in Christian or any other terms) but not to allow reality to other categories of being. "So that it might also take firm hold"—how firm, short of the sign of the Despot? The properly Franciscan rejoinder might be, "Get thee behind me." And do Francis' love and Tristan's suffer a *like* defeat? What, furthermore, of the Franciscan focus on a lower, the Dominican on a higher socioeconomic *class*?

We know that Mumford did a great deal of reading—from the church fathers to Calvin to contemporaries—after *The Culture of Cities* in order to substantiate and refine the formulation of *The Condition of Man*. Yet one is not very sure that the passage on Francis and the desirability of a "firm hold" gains much on the following remark in "Biotechnic Civilization," the penultimate chapter of *The Culture of Cities:* "What Christianity expressed in terms of Heaven, a humanistic socialism expressed in terms of daily living" (p. 378). Reading like Milton's bureaucratic Mammon (spruce up the place in hell, and "what can Heav'n show more?"), such Mumfordian comments are belied by judgments about the "fascist doctrine of totalitarian air warfare: perhaps the deepest degradation of our age" (*CM*, 376) and the "modern world[:] materialistic repletion . . . conspicuous waste . . . organized purposelessness" (*CM*, 380). Perhaps Blake was right; at least Mumford's prophetic *damn*s command more attention than his somewhat relaxed *bless*es. But with the Mumford of these years, intellectual and emotional conflicts, one infers, materialize as generic wobbles in his writing, the teacher to the commonwealth sideslipping into professor of eccentricity, the prophet into crank.

Whether "the Christian" did or did not build "life around the themes of rejection and death" (*CM*, 89; cited above, p. 46), Mumford exhibits in his foreboding wartime book a provocative awareness of death and "death-equivalents"—Robert Jay Lifton's term seems inevitable here. The theme reverberates in a series of contexts: condemnations of illiteracy with regard to the symbol systems of one's culture (*CM*, 9) and dissatisfactions with Apollonian rationality (*CM*, 24–25), with perfected enveloping art (as in the ideal Greek *polis* [*CM*, 30], its proud parochialism, and even its doctrine of the mean, in a life-threatening emergency [CM, 32]), with Epicureanism (CM, 38), and with the aspects of Christianity that he explicates after the "rejection and death" formula.

He portrays the Christian going forth to meet death, partly by drawing a contrast that seems to disparage modern life (CM, 88–89), yet he counters that with an almost condescending view of Christian withdrawal in the Dark Ages:

> Is it strange that in the modern world those who have rejected strong emotions and fancied that life holds no humanly irreparable evils, have also lost all the primitive gestures of grief and have even thrown off the formal costume of mourning?
>
> In withdrawing to a psychological tomb, the Christian treated himself to a second burial, reproducing a condition like that in the mother's womb, when life was in complete equilibrium and held nought beyond bare animation: silence, protection, and peace all recall that primal state of animal unity. If one is not strong enough to fight, one must be discreet enough to pass unnoticed: to lie still and sham death. Grief filled men's hearts everywhere during the long period of violence that broke out in the third century and reached its height, perhaps, in the ninth. (CM, 89)

The restrained deportment of the monk, and his tonsured head, Mumford identifies as a "premature senility" (CM, 93). He surmises in it a design "to make this death-in-life more acceptable." How it should do that he thereupon moots by the complete non sequitur of Eugen Rosenstock-Huessy's thesis about the monks as quasi elders for a society too violent and diseased to provide a normal complement of elders.

Late in World War I, Mumford underwent service as a navy enlisted man. His account in *Sketches from Life* suggests a bipolar experience more common among American men than might be supposed: on the one hand, boot camp and armed service as barbarous; on the other, a fortunate assignment (as a radio operator at Harvard) in a serenely ordered routine of necessary work, and some leisure among uniformed, semitonsured counterparts. It can be a powerful image, both immediately and as a presence in recollection; for me, it meant "retreat" on an army post, with even drivers of cars and swimmers in the Officers' Club pool stopping and emerging to come to attention.[24] Evidently there are always forms of monasticism, and Mumford has been keenly aware

24. During the television reports on Vietnam airstrikes, a colleague recounted how the air force base where he grew up featured not war lovers but the bucolic idyl of fat

that their attractiveness does not necessarily diminish because of the ambiguity of their relationship to the biblical injunction to choose life.

Even holding on, immobile, to an inclusive formulation like Scholastic theology, he argues, is a prelude to death (*CM*, 148). And romantic love such as would, in Donne's words, make one little room an everywhere "appears ultimately as a death wish" (*CM*, 114), but that is a lesser worry to him than the pervasive systems of recent times: utilitarianism in its avoidance of deathly problems by deathly insensibility (*CM*, 307–308), social Darwinism in its misidentifying of the natural with the savage and the life-creating with the death-serving (*CM*, 351), and the sinister master system, the diseased religion of mechanism, by the evil grace of which modern man has been "perverting every hour of life" (*CM*, 380, with homage to Thomas Mann's *The Magic Mountain*) and "painfully committing suicide" (*CM*, 264). He goes beyond Yeats, who complained after World War I of the increasing murderousness of the world, in diagnosing, with Mann, an almost epidemic deathwardness.

It may be objected that *The Condition of Man* is misrepresented, even disfigured, by this florilegium of yew. No doubt in a limited sense it is, as any selective quotation is distorting. But the device can clarify economically Mumford's evident sense that in the midst of life we are in death, and that that is more and more true if our culture is turning (as he feared it is) more and more to the myth of the machine. Furthermore, it can clarify the dynamic and the psychodrama of Mumfordian prophetic surveillance to isolate and concentrate the salient elements.

The chief counterpoise to deathliness in its various guises and modes and orientations is most characteristically adverted to as the "organic." The organic, broadly, is that which promotes life against death most generally. Mumford frequently uses the term in conventional senses stretching from the carbon chemistry of what we *call* the natural world, to the more particular processes of botanical and zoological organs and ecological systems, to the yet more particular interaction and interdependence of organs and systems within the human body. This expectably but problematically opens out from the factors of interdependence and interaction into the figurative uses that imply that the separately

---

sergeants gathered around a soft-drink machine, betting small change on who would get the bottle from the most distant bottling plant.

identifiable elements in culture are as mutually supportive as, say, the cardiovascular and autonomic nervous systems.[25] With the problem of inevitably imperfect metaphoricity, there is the problem of entropy: individual organisms wear out, die, and decay, although organic chemical processes, apart from an immensely distant heat death, may go on.[26] And Mumford debars himself or finds himself internally debarred from any Judeo-Christian conception of life transcending organic chemistry.

But *animal symbolicum* "is to remain fully alive both on the plane of animal existence and on the plane of symbolic participation" (*CM*, 147). Today one may demur at the metaphor of distinct *planes*, even if they are to be understood as intersecting rather than parallel; but this was 1944. And of course, our symbols, in the signs of poems or marble or gilded monuments, may outlast our bodies, as countless mortals since Horace have observed. Since symbol production, or if one prefers, sign production, is nearly as organic an issue of *animal symbolicum* as exhaled breath, and since creation is not ex nihilo, there can be little quarrel with characterizing the coming "back again into life" of relic symbols—as when Shakespeare appropriates Horace—as "the essential systole and diastole of the human heart" (*CM*, 148).

Without following necessarily, it may be true that symbolizations constitute "another transcendental world: a world of durable meanings and values that in time detach themselves from the flux of history, and loose their narrow ties to time and place. Only a small part of human existence actually goes into this other world; and a still smaller fraction is passed on from . . . culture to culture. . . . But that little is infinitely precious; and its accumulation is what constitutes human progress"

25. Some figures culled from many: culture without history like a trickle of water in a desert (*CM*, 13), the organic as roughly "unified" (*CM*, 21, 39), society as a breathing body (*CM*, 73), classic learning as no longer organic (*CM*, 84), the organic Jewish vision (*CM*, 85), Benedictine monastic life like that of an articulated organism (*CM*, 94), fresh cultural combinations as acquiring an organic character (*CM*, 99), Western culture as cancerous and diseased (*CM*, 381, 393), ideas as being ripened for assimilation (*CM*, 418). But compare, just before the clarity and cold bath quoted, the literal and figurative skeletons of knowledge lacking "all that was organic, therefore delicate, transitory, precarious, subtile . . . reduced to something solid, definite, clear" (*CM*, 260). Compare, too, libido as uprush and floodtide (*CM*, 201).

26. Perhaps few readers will want elaboration of the concept of imperfect metaphoricity. But Paul Ricoeur's is elegant, in "Metaphor and the Main Problem of Hermeneutics," *New Literary History*, VI (1974), 95–110.

(*CM*, 148). Given the affective nature and the metaphoricity of *loose
. . . ties* and *accumulation,* one might understand as an act of loving
faith what Mumford appears to judge virtually positive knowledge.

He does not want to be, like the despised Spengler, "caught in the net
of his organic metaphor" (*CM*, 371). What he virtually promulgates,
what he in pioneering fashion suggests and exemplifies in the four gigantic essays *The Renewal of Life* comprises, is a dynamic and symbolically aware interaction with the universe of (as he would say) enduring symbolic creations or (as we might say) texts. Not unlike Francis
Bacon identifying idols of the cave, tribe, marketplace, and theater,
Mumford labors in the four books to characterize referents in Western
civilization and signifieds in Western culture even as they have changed
over time and even as their signifiers have changed noncorrespondingly.
(The word and notion *organic* becomes a kind of adjustable wrench.)

Early, he makes the point that interaction with symbolic creations
could relate very closely to the other mode of necessary interaction, that
of the organic in the sense of botanical and zoological carbon chemistry.
The Benedictines profited from Roman manuscripts on agriculture
when "oral tradition and visible example, in the manors," had been lost
(*CM*, 94). More generally, and more emphatically of culture than of nature, he asserts that rather than "to stereotype a few sorry moments of
the past," we need to be "perpetually re-thinking it, re-evaluating it, re-
living it in the mind . . . deliberately recapturing the past" (*CM*, 12).
That is an antimythic, *biblical* idea of history and historiography, as
Herbert Schneidau reminds us.[27]

Without seeing writing quite in Ongian terms, as a technology that
alters thought, Mumford insists upon the manifold impact upon society
of the writing and printing revolutions. Specified written privileges in
medieval town charters and the like offered deliverance from the "all-
embracing duties" of feudalized agriculture (*CM*, 110). With more
records came a greater fixity and appearance of power, in capital cities.
Mumford likens the emphasis on boundedness, control, and display in
the baroque territorial state to the newly fashionable easel painting
(*CM*, 168–70; *CC*, 79). He virtually sees the capital city as a metatext.
It has certainly, from the days of the scriptoria, housed in significant
buildings the masters—and later the mistresses—of textuality. He

notes, "The process of embodiment no longer demanded the coopera-
tion of all the other arts, as it had in the days of the Cathedral. Paper
becomes the chief medium of embodiment: to become real was to exist
on printed paper" (CM, 255). The baroque artist is a "magnified eye," a
"visual investigator," more than an artisan, as medieval artists had
tended to be (CM, 207).

He sees, does the polytextualist Mumford, that emphasis on the visual and the textual might at a certain degree promote—though *not*
cause[28]—dehumanizing linearity, even powermongering, and likewise
abstraction, atomization, and confusion: "In the newspaper and the en-
cyclopedia the external order of the calendar and the alphabet concealed the conflicts and irrationalities of actual existence: the absence of
true organic relationships in either life or mind" (CM, 262). Here, *true*
organic seems to mean "life-fostering" and to be associated with biological wholeness but also with linkage through the imagination and
emotions to transcendence.

Unconsciously caricaturing the aggregation of file cards, he reads like
an archivist, if not a horoscopist, when he writes,[29] "The subjective and
the objective, the primitive and the cultivated, the mechanical and the
human will finally be united in a new organic whole, which will do justice to the entire nature of man" (CM, 413). We know that *organic*
whole does not mean "compost," nor does *finally* mean the last days;
but how feeble! He can read like a resolution maker, partly reasoning
out, partly intuiting, on the harsh first morning of the new year, a seminal notion: "Balance is not a matter of allotting definite amounts of time
and energy to each segment of life that requires attention. . . . It means
that the whole personality must be constantly *at play*, at least at ready
call, at every moment of its existence and that no part of life should be
segregated from another part, incapable of influencing it or being influ-
enced by it" (CM, 420; my emphasis). Clearly the "negative way" is not
for him nor by his lights for us; like Milton, he can praise no "fugitive
and cloistered virtue." One wonders if he thinks of serenity or tranquillity. More broadly than Milton, who at that place in "Areopagitica"

28. Lasch argues Mumford's antideterminism well. See especially "Mumford and the
Myth of the Machine," 11–12.

29. Lasch notes and acknowledges the partial justice of some negative past criticism.
Wilfred M. McClay seems to me much too sweepingly negative about "this would-be
Jonah . . . *philosophe*" ("Lewis Mumford: From the Belly of the Whale," *American*
Scholar, LVII [1988], 111–18).

conceives a "race" for the "immortal garland" of salvation, Mumford seems to suggest a sort of internal and perhaps external team field sport. Or even free play? The bibliography of *The Condition of Man* lists three books by Johan Huizinga, but *Homo Ludens* is not mentioned until *The Conduct of Life*, in 1951. Perhaps more important, Mumford's words and spirit seem to anticipate in a downright American way constructions that later came ramified from France—the "at ready call" in the *disponibilité* of Gabriel Marcel, the spirit of strenuous play in the *jouissance* of Roland Barthes and others.

Mumford writes like a latter-day version of Western man as pilgrim when, in the same section, "The Organic Person," he sides with Jakob Burckhardt against the "terrible simplifiers" and avers that "we must . . . find a benign method of simplification. . . . The task for our age is to decentralize power in all its manifestations" (*CM*, 419). Rather more generally and pervasively, he seems to imply that we organic symbol makers need some community with others transcending ordinary human failings, and some antiphony within, which might harmonize ordinary human discords. He comes at last to declare, "The deepest, the most organic, of these higher needs is that for love" (*CM*, 413).[30] In all that, he writes as no hypocrite lecteur, [mais] mon semblable, mon frère. Certainly he wrote to himself, "There will be many fourth volumes, if the change in attitude I have tried to make possible, first of all in my own life, really takes hold. . . . Blissful is the word for these days" (*VWB*, June 28, 1944).

Though I reserved for *The Conduct of Life*—and my own further maturity—discussion of the final problems of man's nature, destiny and purpose, the present volume, so far from being an epilogue, is in fact a preface to the earlier books. While each volume stands alone, they modify each other; and the full import of any one cannot be grasped without an understanding of the other three.

30. Mumford poses a fulfillment scale against a survival-to-personal-growth scale that resembles that given currency by the late psychiatrist Abraham Maslow. See below, 65*n*37.

So, in the spring of 1951, Mumford wrote in a prefatory note printed at the front of *The Conduct of Life* but listed near the end of the table of contents, as if itself a model of the book-length epilogue become preface. Anyone who has written the introduction to a doctoral dissertation or a book only after writing the rest of it may experience a surge of fellow feeling: we come at last to know better what we were doing all along. But the conflicts he testifies to in the composition of, and manifests in the text of, *The Conduct of Life* go beyond that; and their implications for the four volumes of *The Renewal of Life* both resist and reward disentangling.

The struggles and delays seem to have involved self, others, and subject—together, and by turns. Or to put it in an idiom much less his own, he had to struggle with issues of identity, signification, and the rhetoric of intersubjectivity. The struggle seems to have been a prerequisite to the historical triumph of *The City in History* and the prophetic triumph of *The Myth of the Machine;* the struggle was interesting in itself as exemplifying a wisely critical but troubled American in conflicted times, and it afforded successes of occasion and degree but also lapses and missteps. So he recognized when he testified to the "critical acumen" of Paul Rosenfeld:

> This applied equally to my own work. When I sent him "The Condition of Man" in 1944, I eagerly awaited his criticism and hoped for some measure of praise; but months went by before I heard from him, for, he confessed, he found writing me difficult, since my book delighted and yet dissatisfied him. "I felt its point of view right," he said, "but ineffectual, since it made a moral issue of the matter of regeneration; and I felt that all movement comes from connection with a new object, or a new connection with an old, and found yourself taking to what I feel to be a moral exhortation. . . . All that you say is true. But a constatation [authentic statement] is not necessarily an act." That criticism was more salutary than any praise; and though I took it to heart slowly, the more I considered it, the more dissatisfied I grew with the final chapter of that book. Unfortunately, this did not keep me from repeating the same error even more blatantly in "The Conduct of Life." (*SfL,* 375; Mumford's brackets)

The questions of what condition and bearing of self would be requisite to address some such book (and to what audience? in what language?) were alive to Mumford early as well as late. In a long letter to

Christiana Morgan, on August 4, 1944, he came around to a somewhat dancy (defensive? puzzled? quasi-bluff?) backing and filling:

> I deliberately postponed *The Condition of Man* till I reached the fulness of my maturity and had felt at least the first autumn frosts of old age. . . . By now I feel impelled to quicken the pace of my writing, lest I get too far away from the heat and passion out of which whatever wisdom I now have has been distilled. For the weakness of most moralists, philosophers, and religious teachers is that they draw mostly negative conclusions from these data; whereas I want, rather, to indicate the positive incentives that previous systems have either left out or attempted too prudently to disguise.
>
> My present tranquility and peace may be due, not to wisdom and balance, but mere hormonic inactivity. (*W&D*, 182–83)

Things unattempted yet in prose or rhyme? "There was a moment" when he thought he "could handle *The Conduct of Life* only as a poem!" (*VWB*, January 22, 1952). The need to regard traditional morality as life-denying, and Christianity as especially so, was not unusual for him, as we know. One wonders at the need to deny Milton's "Hail, wedded love" and Shakespeare's treatment of the blessedness, whether Christian or not, of romantic and familial love. One may note briefly what kind of social or moral vision his self-defined position as maverick fostered and consider his need to claim major novelty.

The poignantly ironic letter to Brooks written out of "blissful" days (see above, p. 54) even more explicitly connects autobiographical ambivalences and conflictedness of rhetorical situation:

> As for *The Condition of Man*, you are entirely right in your diagnosis. This is the first part of the final volume; and Volume IV, the most difficult job of all, still lies before me. The present book ran away with me, despite my different plans for it; and I think it did this for a good reason: there is something in our long common past that we need to reason with and struggle with if we are to be strong enough to build the future. We must enter purgatory before we dare to visualize heaven. . . . With this book written, I could now die happily; because I think that once its message is digested, the reader could write his own fourth volume; indeed, there will be many fourth volumes, if the change in attitude I have tried to make possible, *first of all in my own life*, really takes hold. . . . It will take two or three years of the hardest kind of concentration to write Volume IV; and though time presses on us all, I have no intention of undertaking the final task till I am again in the pink. (*VWB*, June 28, 1944; my emphasis)

The Mumfords learned in October of the death of their son, Geddes, a month earlier in the Italian campaign. Letters to Frederic J. Osborn, his friend and ally in the cause of town planning, attest poignantly to the profundity of his distress and dislocation: "After a two year interval of idleness—I was once more absorbed in my work" (*FJO*, August 12, 1946); "By writing like a madman for three weeks . . . I managed to do the second draft of my memoir on my son" (*FJO*, October 10, 1946); "After [April] I shall get to work in good earnest on volume IV, . . . practically the foundation of all my other works, and my last will and testament. After *that* the only graceful thing to do will be to die promptly . . . before I become an embittered old man" (*FJO*, January 1, 1947). In that last, one may be tempted to discount the New Year's Day stocktaking and resignation as verging on the playful. But Mumford, who has championed the play element in culture, exhibits little playfulness in his many published letters. And one sees that the work that absorbed him was first to memorialize his son in a book (published in 1947 as *Green Memories*) and then, "in a mood of Melvillean bitterness" for himself as a "pariah, except out of my own country" (*VWB*, April 14, 1946, November 13, 1947), to memorialize himself as if the life of this internal "exile" were over.

The leaden death that depression can be, the anguished mood swings, the internal pitch and toss, the gradual reacceptance and reassertion of life, of life in more stable vein—one recognizes the grief syndrome working itself out, as we sometimes say. But of course *out* does not mean "into some pristine state." Mumford wrote at length to Osborn on Christmas, of the two of them being "both, so to say, children of Wells and Shaw . . . for myself, likewise Chesterton." He complains that those writers were sometimes a "very bad influence on my sexual and marital life, right into maturity: particularly Shaw, with his cool rationalism and his cerebral sort of sexuality." If feeling "betrayed by Shaw," he continues revealingly that he has outgrown the inadequacies of Wells, among them "his over-reliance on science and education: he never realized that there were depths to man's character that needed to be *touched* before these could have their desired effect. All that religion describes as '*grace*' and *conversion* was absent from his outlook" (*FJO*, December 25, 1948; my emphasis). Although Mumford's quotation marks around *grace* betoken some ambivalence, he clearly implies a measure of conversion in himself and his vocation, which glosses his claim that the fourth volume somewhat rearticulates and reorients retroactively the work of which it is a part. And he goes on to do more than imply: "But [Wells] was only half a guide; and one of my own missions has been to give to my younger generation some of the dimensions of human experience that both these great figures left out of their vision of life." So, as he was working on this "foundation," this "last will and testament," he was working to compensate, working to be the better father than his absent and surrogate fathers had been, working to be the completer father than war had allowed him to be to this son. Writing *Green Memories* had obliged him, now sonless, to think a good deal about parenthood and his own oddly fatherless status, and even of the language of fathers.[31]

Well along in *The Conduct of Life,* he wrote two long letters that illuminate the conflicts in his position with respect to himself and some or another audience and with respect to the very terms, the signifiers, of the discourse. Alone in the house in Amenia, New York, during the week, while his wife and daughter were in the city, he complained that he formerly loved such solitude, but "somehow my son's death seems to have changed this feeling; . . . one ear is always cocked . . . for a footfall . . . that never sounds." As to the book, it

> is one in which I deal, from beginning to end, with the eternal platitudes; and it would be sickeningly impossible to read did I not embellish my conclusions with such modifications as my actual experience of life suggests: before each monumental truth, respected for ages, I have as it were to plant the tiny living sapling of my own experience, which will cast its shadow on the monument. To do this with a certain degree of honesty will be the test of my abilities, not as a writer, but as a man. In the chapter on marriage and the family, for example, I must say all that I believe and actually live by on that subject. . . .
>
> But this is not the time for me to be telling you about the book; except that I've had the old experience of finding that the hard parts, about religion and love, have come relatively easy, and that the easy parts, on which I am now working, those dealing with economic and political facts, bringing forth arguments and principles that I know by heart, have proved to be the hard parts. (*FJO*, June 9, 1949)

31. In *Sketches from Life,* Mumford tells the remarkable story of an émigré supposed father, of an instructive, attentive, kindly, but *old* foster grandfather, and of the surmise growing with his adult years but verified only in 1942 that his biological father was a lover his mother refused to marry.

Somehow changed? Why should there be anything puzzling in the house of absence unsettling him? It was no doubt a harsh lesson to learn at fifty-four. Readers of the book in its published form may feel that the hard parts about religion and love come rather too easily to Mumford to engage us satisfactorily and that the lived experience sometimes appears less as graceful shadows on the monuments of truth or even as embellishments than as washes of assertiveness leached out of observations and events uncertainly conclusive. Sickening platitudes? That is probably in part Mumford's displaced rage at the puzzling hurtfulness of life. More surely it is impatience with taxonomizing, as his reiterated word *experience* suggests. For the generic expectations of personal history, the book is both too autobiographical and not autobiographical enough; for the generic expectations of philosophical reflection, it can seem too long in assertiveness and too short in argument. Three months later he wrote to the writer David Liebovitz, a friend from navy days in World War I:

I am still not satisfied with it; and contrary to my original resolution, or at least hope, I'll probably have to take another lick at it this winter, before I turn it in to Harcourt's. If it hadn't been for Geddes' death and the general drought that came into my life the last five years, I'd probably have gone on writing this book as soon as the Condition of Man was out of the way; and it would have been carried on in the great tide of energy that, despite all backward washes from the war, carried me through on that enterprise. The present book will lose something by the hiatus, something which my present health and energy, somehow, does not make up for. I was an ill man when I wrote The Condition of Man, but the book itself shows spiritual health and élan. As to the present manuscript, I am not so sure: the writing of it has been a tremendous act of the will, but without the inner buoyancy, or rather the inner pressure, that kept me at the earlier book. Some of my present difficulty comes from the fact that I am overprepared for the present work, have lived too long with the material, and am bored with what may be to others very fresh and even sometimes original thoughts. As to that, one can't tell. . . . It would be inconvenient, if not disastrous, to hold up the publication. . . . (I gave The Condition of Man a year for reflection, and though that led to no very drastic revisions, it probably benefited the final chapters and improved a lot of little things.)[32]

32. Bettina Liebowitz Knapp, ed., *The Lewis Mumford/David Liebovitz Letters* (Troy, N.Y., 1983), September 9, 1949.

He was right, of course: the reconfiguration of a more mature self in a less disjointed relationship would take more than a year, was taking more than the two decades of *The Renewal of Life*'s four volumes. The remarkable ambivalence and conflictedness indicated in the letters may be seen partly as an index of the difficulty in reconciling—or alternatively, choosing between—what Rosenfeld had called constatation and act.

The conflicts tended to become acute in connection with the question of how to conclude each volume. The almost obsequiously admiring and supportive Brooks opined in a preliminary response that "you have opened up a new vein that calls for a further development in many more books. I am thinking especially of 'The Way and the Life' at the end of this book" (*VWB*, September 13, 1951). Mumford quickly responded with thanks for the praise in the letter, and with explanation verging on defensiveness in response to the not quite complete praise Brooks had offered: "You are right in thinking the last chapter only a beginning: what follows from it are the three chapters of *application* to education, to politics, to marriage that I finally chucked out of the ms. as too meager to do justice to the theme, yet so bulky that they would, if left in, sink the book" (*VWB*, September 18, 1951).

It is worth following the dialogue farther, not only for what it reveals of *The Conduct of Life* and for the conflicts it signals within the American moralist Mumford but also for what it reveals by contrast with the stance and sureness of the writer's great later work.

The *New Yorker* of November 24, 1951, featured a substantial review by Lionel Trilling of *The Conduct of Life*. Mumford immediately drew this to Brooks's attention, and he returned to it plaintively in a very long letter of rejoinder to Brooks two months later. Trilling was at pains to be fair to what, as Mumford recognized, he found a highly *decent* book quite knowledgeably and generously grounded in a very extensive Western tradition of moral writings.

Trilling's fundamental complaint, a shrewdly incisive one, is that Mumford's "method of synthesis" is unrealistic both in what seem to the reviewer relatively small matters, like the call for a synthetic international language, and in large ones:

> He will not, for example, believe in a God who is at once the omnipotent creator of the universe and also its benevolent governor, for human suffering makes this an unacceptable paradox. He therefore recommends belief in a God who is incomplete but always evolving in our conception of him and

whose full realization will be . . . "the very last stage of all observable development." At the same time, he wants all the spiritual benefits of Western religion, the sense of transcendence and mystery, of universal purpose and drama of evil and sin, not stopping to consider that these have been developed through the acceptance and contemplation of the theological paradox he rejects.

The "impulse to synthetic perfection" makes the book "intellectually unsatisfying" and "perhaps worse, . . . politically confusing," according to Trilling. He magnanimously refrains from likening Mumford's God in *The Conduct of Life* (indeed, throughout *The Renewal of Life*) to Mark Twain's father ("But when I was twenty-one, I was amazed at how much the old man had learned in four years"). Nor does he anticipate Eric Voegelin's excoriation of Gnostic and neo-Gnostic "killings" of God by arrogating divine creativity to man.[33] Trilling's summarizing is superb. On the other hand, his complaint against the pursuit of a logically and psychologically impossible synthetic perfection may lose some of its force if one thinks not of impossible, magic resolutions within individual personalities but rather of the playful dance of incongruities Mumford seemed to endorse, and if one thinks of societies not as homogenized but as syncretistic, with the mixed governance he applauded.[34] Mumford was a killer neither of God nor of men, not at his most bitter.

Mumford in his long letter of January 22, 1952, complained that Trilling "overstressed points of resemblance to the Victorian writers"; one may note that a reader somewhat oriented, like Trilling, toward nineteenth-century writers might well fasten on Mumford's endorsement (repeated in *The Condition of Man* as in *The Conduct of Life*) of Ruskin's assertion that there is no wealth but life.[35] Trilling and Brooks both "overlooked what is fresh and original," Mumford believed. But

33. See especially Eric Voegelin, *Science, Politics, and Gnosticism* (Chicago, 1968; first given as a lecture at the University of Munich, in November, 1959), and Voegelin, *The New Science of Politics: An Introduction* (Chicago, 1952; first given as Walgreen Lectures at the University of Chicago, 1951), Chapter 4.

34. Compare Mumford's letter to Babette Deutsch, May 31, 1936: "of oneself . . . we are a federation" (*Findings and Keepings* [New York, 1975], 359).

35. From *Munera Pulveris*. Michael A. Weinstein argues, in his fascinating *The Wilderness and the City: American Classical Philosophy as a Moral Quest* (Amherst, Mass., 1982), that "arguments promoting moral commitment [and] moral community . . . are the soundest elements of the American tradition" (p. 29). Compare Brooks: "It carries on the major American line of Emerson, Whitman and William James" (*VWB*, September 13, 1951).

Trilling seemed to misprize rather than overlook the novelty. Brooks thought it new for Mumford to call for tenderness (*VWB*, September 13, 1951)—which Mumford did in a way that was not gender-specific. Had Brooks himself not been so immersed in nineteenth-century American writings, the kinship, whether influential or not, of eighteenth-century "sensibility" to what Mumford commended might well have seemed significant; still, tenderness was no conspicuous topic in American discourse in the decade ending in 1951, unless ascription to stereotypical femininity. Mumford did some American trailblazing there.

It remains to decide whether he was a pathbreaker, as he thought, in something of a larger, more generic sense:

> I addressed the [four books of the] series to those who had some learning and "background" in the field covered . . . who yet lacked certain elements of humanist discipline that I . . . could bring to the subject. . . . But [the four books] resemble more closely a new kind of writing that grew up first in the nineteenth century: a species represented by *Moby-Dick* and by William James's *Psychology*, in which the imaginative and subjective part is counterbalanced by an equal interest in the objective, the eternal, the scientifically apprehended. This is no shallow fashion; and my recourse to science is not merely an attempt to make people believe what I say, by backing it up with scientific statements, as you suggest. Far from it. What lies behind the new method is the conviction that personal experience and personal intuitions, however sound, do not carry full weight until every effort has been made to square them with what William James called "hard, irreducible facts," meaning by that, results that have proved valid for all sorts of different observers, in consequence of the application of the scientific method. . . . But this combination of personality and individuality with impersonality and collective research is what, it seems to me, characterizes the best thought of our time. (*VWB*, January 22, 1952)

What *balance* or *counter-balanced* can mean here or throughout *The Conduct of Life* and the series is questionable. There is here, as frequently, a hint of reductive materializing and spatializing—as too in *square* and James's *hard facts*. Mumford's interpretation of the supposedly equivalent notions of proof, validity, and scientific method may in this context represent insufficient discrimination between shareability in the communicational sense and repeatability in the scientific sense. For collective research may be less idiosyncratic but not less personal than individual research or reflection, though it may be more obscurely and indirectly personal. Can we—does Mumford mean that we—read his uneasy interaction of the personal and the impersonal as *allegory?* His bold generic experiment . . .

Mumford's follow-up letter to Brooks refines somewhat the sketch of the writer's intellectual lineage and of the generic posture in *The Renewal of Life*, and it can help differentiate that earlier work from the great later work:

> If my thoughts are to have some immediate effect upon my contemporaries, instead of seeping down slowly from the few to the many, they must be written in a simpler style. . . . My aim has been to develop a style that would be supple and ready for use on the greatest variety of occasions [and] readable over long stretches, by reasons of variations in tempo. . . . The man who perhaps comes nearest to my ideal of style is Jean Jacques Rousseau—though I cannot bear his style in his more maudlin sentimental moments for the exact reason that I cannot bear the sentiment itself. (*VWB*, January 26, 1952)

Rousseau! What is one to say? Although Mumford read French, one does not suppose—and his animadversion about sentimentality does not suggest—that he means any very specific equivalence of diction or syntax. His aim, and evidently his achievement, have been something more generally akin to the rather straightforward-seeming account in *The Confessions* and *Emile*, and to the prescription in *Emile*. Whether Mumford is as much the quasi confessor as Rousseau is a question worth keeping in mind. But surely he follows Rousseau in acknowledging that even laboratory observations must come from the observer, however helped by instruments, and not simply from the instruments (including the scientific method).

In any case, now more than a generation after the writing of *The Conduct of Life*, what is likely to interest the student of contemporary culture is Mumford's struggle to realign incongruous signs and referents, even signifiers and intractable signifieds, with regard to the conduct of life in a life world seen as radically conflicted and temporally overdynamic both in history and personal development. It may be almost emptily inconclusive to stipulate that the "social breakdown of our time" shows itself "philosophically, ethically and politically" (*CL,* 149); the lines of signifiers do not convincingly mark joints in what may be a monolith, or else they mean too many different things to different readers. Perhaps it is truistic to assert a moment later that the degradation of values carries meanings with it "toward nihilism"; but it is suggestive to adduce "Nazi anthropology, Aryan physics, Stalinist sciences," which, although Mumford does not quite yet put it so, are later glossed as tropes of power, or more exactly, as tropes of the displacement of power—as virtual allegories.

Mumford's three criteria for an "ethics of human development"—as usual, in the dual sense of individual and cultural development—are a reverence for life (with a deep bow to Albert Schweitzer), the development of evaluation and selection (never much less vague than that), and an acknowledgment of the purposive nature of all living processes (with special attention to goals to be pursued over a longer term than the individual human life; CL, 139). The third criterion may be extremely open, but it seems more provocative of discussion and investigation at levels from the biological to the theology of play than the second criterion.

Mumford was then giving heavy credence to Dr. William H. Sheldon's notion—much bruited in the forties and early fifties—that personality is differentiated by types in embryonic development: the endomorphic, mesomorphic, and ectomorphic body types. With some debt to the more or less analogous distinctions drawn earlier by Auguste Comte, Patrick Geddes, and Victor Branford, Mumford urges important correlations between the endomorphic privileging of internal organs and the stage of "Dionysian" youth, the mesomorphic privileging of the muscular-skeletal structure and "Promethean" maturity, and the ectomorphic privileging of the nervous system and "Buddhist" age (CL, 193–200).[36]

That this neglects innumerable questions is obvious. What of all those skinny tennis and basketball players? aging powermongers? anorectic women? Mumford takes *some* correlation of body and personality to be evident. That his own associations are driven partly by a strongly reductive view of Judeo-Christian history is patent: the "failure of the universal religions, perhaps," has been that each "bore too plainly the mark of a simple biological type: the cerebral" (CL, 195). Moses the reclusive Talmudist? Amos the anchorite? Jesus returned from forty days in the wilderness to a long life of reflection (in the temple, with the moneylenders)? It is as if some drying-up in himself that his letters allude to was reflected in a hypostatic tendency of his signs that is sometimes al-

36. This leads, for one example, to a silly association of the venerable Western idea of all life as pilgrimage with considerations about the remotest "trajectory"—"almost the invalid's level"—of advancing age (CL, 199).

most allegorical, as here, and also a few pages later: "No historic religion has yet sought to sustain life in its fullness and wholeness" (*CL*, 200). This is also a scarcely disguised cry of bereavement.

Closer to the street and square in *The City in History*, closer to Adamic "right names" for what might better be called "leading irreligions" in *The Myth of the Machine*, Mumford partly identifies the problem of *The Conduct of Life* and partly so dramatizes it as to make the book's title ironic in somewhat the way *The Education of Henry Adams*, as a title, is ironic—and valuable in some analogous ways. Yet his reflective roughhewings are more shapely than the abstractions of intellectual fad and ideology that were widely current at the time he wrote. Mumford honored human development, construing it in terms of the successive dominance of situation, talents, and time of life, and conceiving it as both interactive and intraactive. (Thus, he updated Taine's *race, moment,* and *milieux,* with perhaps a *fumet* of cycles and epicycles.) The scheme is crude enough. Does one say situation or time of life dominates in infancy and senescent age? Does not the situation of serious illness, surgery, or injury dominate at any stage of life or level of talent? Does not even a temporary temperature of 103 dominate? Are not inside and outside both problematical and yet unavoidable concepts with regard to the self? Still, Mumford's filing-cabinet taxonomy, by its diversity and antiexclusionist ethic, invites more analytic richness than the bipolar arguments then current in discourse about heredity and environment. And his growing sense of developmental dynamics can lead farther from naïve and strident positivism. He seems not to have known Abraham Maslow's model of a developmental hierarchy of human needs (those of physiological safety and security, affection and belonging, esteem, and self-actualization).[37]

Language, whether Rousseauist or not, and dynamism go together. A familiar reason for that is that all the resources of language are needed to bring the horizon of human potentiality and significance into intersubjectivity and discourse. If Mumford in the late 1940s was not prepared to put the matter quite in those terms, if indeed some positivistic terminology and inclination lingered with him, still his increasingly inclusive disposition could show itself well enough: "The various contem-

37. See Abraham Maslow, "A Theory of Human Motivation," *Psychological Review,* L (1943), 370–98, and Maslow's subsequent writings. Compare Mumford's proposal of a fulfillment scale (*CM*, 413).

porary reactions against the full employment of language, from da-
daism to logical positivism, will not in the least save us from error and
self-deception: they merely substitute for the small detectable errors of
misused speech the colossal error of rejecting the greater part of man's
subjectivity, because it comes to us primarily in symbols of a non-
operational order. . . . There are no purely physiological processes. . . .
When one begins by defacing the word one ends by defaming life" (*CL*,
52–53). It might be argued that this rests on a discredited notion of
some transcendent yet available objective standard against which
"error" unequivocally appears. But, on the whole, despite Mumford's
opinionated moments early and late, his writerly conduct appears to endorse exactly the complementary notion: that language obtains in a universe whose deity he quasi-conceives as a God-becoming, that language
is vulnerable (like life) partly because always a matter of sharing and
negotiation within (and between) language communities. Three chapters later, he writes, "The first step toward freedom will be a new respect for the symbol" (*CL*, 144). It is a profession of faith in the possibilities of recognitions and of contingent but approximately shareable
relationships of signifier and signified and, similarly, of signs and referents.

Although *sources* and *influences*—however problematical those
terms may be taken to be—are not the preoccupation of this study,
Mumford's bibliographic annotation on *Toward a Philosophy of History* is notable: "Ortega's approach to history is similar to that of this
series" (*CL*, 309). If as retrospect that may not fit the first two volumes
as well as the latter two, it is fair enough as prospect. Ortegan history
and the lived-in forest of *Meditations on Quixote*, with disclaimers of
idealism *and* realism, are near at hand when Mumford takes issue with
Hannah Arendt a few years later in one of his "Random Notes":

> Hannah Arendt's conviction, similar to Vico's, that man can know only what
> he creates, should have a different conclusion than that which she comes to.
> On her reading the real world is closed to man: what he makes of it is noth-
> ing more than a fresh picture of himself, with his limitations. What she for-
> gets is that man himself is not extraneous to the universe: he is part and par-
> cel of it, and so the self he seems to project is in fact an integral part of what
> he investigates. Therefore man's picture of the world, instead of being con-
> ditioned by him, is the very world that he seeks. Through his own personal
> and communal activities the inner and outer become demonstrably one.
> (March 20, 1967, in *W&D*, 192)

So much for Gnosis! He would not by then be nor have others be (in Eric Voegelin's sense) adepts or power agents of a secret knowledge conceived to kill God and reconstitute a vile world in some merely human perfection. Perhaps in *The Renewal of Life* only his sensitive feeling for complexity and his love for the world immunized him.

This quotation and the one from 1951 on language defacement and "defaming life" (see above, p. 66) help fix in a useful way Mumford's otherwise wayward-seeming employment of the term *organic* and of expressions closely associated with it. He often uses the term in contexts that appear to attach it to familiar, conventional, nonmetaphorical botanical and zoological referents (*e.g.*, *CL*, 9, 14, 24, 27, 30, 31, 33, 37, 128, 132). Occasionally, of physical events in nature and their relationship, the term will seem to mean what has come ordinarily to be called ecological (*CL*, 23–24, 140–44, 223). Mumford's highlighting of the *integratedness* of life processes—as if in one living body—tends toward marked degrees of metaphoricity, as, for example, when he writes of ideology (*CL*, 23), of the value in life as organic mixture (*CL*, 126), and of organic knowledge of the self (*CL*, 247). There is a similar metaphoricity in his highlighting of the *dynamism* of the living body and the life process when he describes the organic overflow in vocal play (*CL*, 42), the tiny "seed" of personality in a culture (*CL*, 108), the cancer cells in the body politic (*CL*, 115), the organic compost of discrepancies in social systems (*CL*, 176), the slow organic procedures (in contrast to Marxian struggle; *CL*, 203), and the attempt "to think organically, that is, simultaneously at every level" (*CL*, 261).

One would not willingly whore after false unities, nor blink discrepancies. But I believe we must recognize genuinely unifying power in Mumford's disposition to base himself on a position, neither idealist nor realist, to which the phrase "the mind as nature" points, a phrase Eiseley was to formulate some years later when he and Mumford were colleagues at the University of Pennsylvania. Mumford recognizes, like Eiseley, that nature and culture are not neat or symmetrical complementarities, even when he writes, "Each is inconceivable except in terms of the other" (*CL*, 39; see pp. 34–35 for "nature" as, *inter alia*, "poet" and "mechanical engineer"). He resembles Eiseley, too, in the widespectrum inclusivity of his sense of interaction and dynamism. In that "nature" which, in virtue of our eyes, can in an obvious though too simple sense be said to lie outside mankind, both men find (in Mumford's word) the "exuberance of life . . . [that makes] all our rational standards of economy seem mean and restrictive" (*CL,* 34). Both men, with regard to the world of nature and culture we call "within," honor dream, and they honor it similarly. For Mumford it is the liberator from neurotic routines and the experimenter (*CL,* 47), a principle of life for any civilization (*CL,* 49), the honorific complement to remembering (*CL,* 53), and the superior antecedent to mere tool using (*CL,* 50).

I have altered Mumford's expository order in listing these citations because he approvingly acknowledges Carlyle's phrase *man the tool-user* but takes issue with Bergson's redefinition of man as maker more than knower, *Homo faber* more than *Homo sapiens* (*CL,* 40–43). Bergson, whose emphasis on the mode of time as lived experience, as duration, was congenial to Mumford, accordingly seemed surprisingly—almost perversely—wrong to identify intelligence with "the geometrical, the mechanical, the non-living," in a word, with the static or abstract.[38]

This point can lead to some of Mumford's language for drawing contrasts to his ideal of multielemented, interactive dynamism. He recognizes the mechanization (to speak metaphorically) of the doings of everyday life as corrupt reductions of "routine and habit" (*CL,* 173). Similarly, he laments dehumanizing amplifications, extrapolations, and fetishizings of the potentially benign gifts of technology, such as the "god of Speed," to which the "American people sacrifice more than thirty thousand victims every year" (*CL,* 82). The reductive abuses of design and dynamism can by their very oppressiveness beget angry and destructive responses, he observes in a paragraph on the appeal of nihilism. But the truly framing contrasts, the genuinely defining Other to the organic, emerge from Mumford's implicit definition of the organic as life with mysterious complexity fostering life: the Other is thus hatred that would reduce its surroundings to manageable crudity or even to death. A crucial part of Mumford's organizing sense of Western culture in the final two volumes of *The Renewal of Life,* in *The City in History,* and in *The Myth of the Machine* is a deepening anxiety over a resurgent hatred of life and over reductive dismissals of complexity: "The violence and evil of our time have been, when viewed collectively, the work of loveless men" (*CL,* 283). Is not this point, central to the life

38. For an interesting appraisal of Bergson's important place on the contemporary American intellectual scene, see Paul Douglass, *Bergson, Eliot, and American Literature* (Lexington, Ky., 1986).

world of Mumford's convictions, similarly central to the worlds we inhabit? "Our fear of emotions, our habit of treating normal emotions as deplorably sentimental and strong emotions as simply hysterical or funny, betrays fundamentally our fear of life" (CL, 153). *Our* may be annoyingly vague or seem unwarrantedly general. But then it is possible to observe, in discourse at large, how widely current is the perhaps defensive pseudodichotomy of the cognitive and the affective. Indeed, each popular advertisement heard, seen, or read will implicitly fortify that defensive idea, each advertisement that builds on the stereotype of the machinelike working professional, whether or not it draws a portrait in which there is a contrast to a mindless, beery, disconnected time of play. Edmund Muskie's political candidacy in 1972, crippled by his angry tears in New Hampshire, comes to mind too as an emblem. Mumford evidently intends awareness of death as a valuable defining contrast but not precisely a binary one; life and death are not to be conceived like a mechanical on-off switch. Values, as a central instance, arise "like a coral reef" out of "millions of lives that have never quite passed away" (CL, 26). Man, Mumford adds, "paid for his creative exuberance by his increased consciousness of death . . . a new dimension to his life" (CL, 36). Accordingly, "our withdrawal of interest from death, in itself points to an erosion of values" (CL, 127; again the ambiguous *our*).

His point was bold for its time, albeit roughly contemporaneous with Evelyn Waugh's *The Loved One* and Jessica Mitford's *The American Way of Death*. It may tend for today's reader to meld into the modestly resurgent acknowledgment of death signaled by such writer-teachers as Elisabeth Kubler-Ross and Robert Jay Lifton and played out in newspaper and television features, minicourses in modish suburban high schools, and the like. Mumford's point is at least twofold. One part of it is that denying or hiding or cosmeticizing death obscures the vulnerability of life. And "in the highest realm of all—the realm of personality—life is even more delicately poised over the abyss of non-existence than it is on the organic level" (CL, 33). But, as his analogy of the coral reef indicates, death and nonexistence correlate very imperfectly, and presence and absence as lived experience do not complement each other in so polar a fashion as those signifiers suggest. To ignore or psychologically to deny death risks an intolerable loss of history, loss of those millions of whom he spoke who can persist as effects and meanings present to diverse minds. Life is vulnerable, yet paradoxically strong if we *choose* to perpetuate the meanings—which must include the deaths—of those millions.

The dynamic, the life-affirming, ultimately the loving, features of the variously organic are spelled out by Mumford in multiple ways—psychologically, biologically, existentially—by drawing contrasts with static formulations, mechanistic constructions, and a hatred of life. This is to amplify what he wrote: "Continuity: emergence: creativity—these are the basic postulates of the new synthesis" (*CL*, 25). By implication, discontinuity, regression, and stasis or destruction characterize the old catalysis, or at least that past which must be, should we say metabolized? and transcended.

The good and bad triads ramify as concrete and lived experience. They also ramify at large in *The Conduct of Life*, and may be sketched in abbreviated form here, as a kind of psychomachia: as an inchoate sense of transcendence against useful but stultifying epistemology and ecclesiastical and quasi-ecclesiastical systems.[39] Mumford writes that "the scepticism of systems is a basic thesis of this book, but it has another name, the affirmation of life" (*CL*, 177; see also pp. 178–79; acknowledgment of debt to Patrick Geddes and Charles Horton Cooley, *SfL*, 158).

Psychomachia would account for some of the weariness Mumford reported in connection with writing the book, weariness unaccounted for if one notices only assertiveness. Most generally: "None of the existing categories of philosophy [or] science or religion [or] social action, covers the method and outlook presented here. . . . [This is] the philosophy of the open synthesis" (*CL*, 179, 180). And in crucial particular: "In all humility the present philosophy affirms as persistent that which every system of revelation tends to coyly modify or arrogantly deny: the continued existence of mystery itself" (*CL*, 88; similarly, p. 57).

He risks self-parody in the strident humility, the tendentious metaphorics of *cover* and *open* (as of a shrunken blanket versus the great outdoors), and the apparent blind eye he turned to the awe many other outsiders have judged to empower Islam and Buddhism and have in

39. I have intended the religious associations of such language since hearing Eric Voegelin lecture on neo-Gnosticism over twenty-five years ago and reading Ernst Cassirer's *The Myth of the State* (New Haven, 1946). I am delighted to find closely argued corroboration in Stephen A. McKnight's *Sacralizing the Secular: The Renaissance Origins of Modernity* (Baton Rouge, 1989).

America been able to descry easily in post-Reformation Christian eccle-
siology and theology. But, apparently, his seeing eye is on a social-
gospel Christianity (*CL,* 170–72) or an otherworldly, feeble church of
pale Galileans, of "smiling accommodation rather than resistance" (*CL,*
101–102; see also p. 63). That he finds fault with the American Judeo-
Christian establishment for being too worldly and for being too un-
worldly was not unfair or inconsistent: the diversity in that unwieldy
structure certainly permits local spokesmen and congregations to represent extremes that can be abusive and exclusive.

Mumford evenhandedly criticizes, on similar grounds of exclusivity,
humanistic reductions of evil to mere error (*e.g., CL,* 151) and positivistic "omissions" of purpose, value, free will, potentiality, ideal, and final
goal (*CL,* 58–59). And exclusivity is the central issue in Mumford's
theological position. It is not quite that he wants a God without awesome
paradox, as Trilling supposes; it is—even if unwittingly—a more orthodox point that he makes: "The universe does not issue out of God, in
conformity with his fiat: it is rather God who in the long processes of
time emerges from the universe. . . . God exists, not at the beginning,
but at the end" (*CL,* 71). This is neither to displace divinity to "processes of time" nor to complain with Nietzsche that God is dead or not
yet alive. It is more nearly an orthodox complaint about the sign for
God. As has been said of Nietzsche's cultural milieu, that "God" was
dead for Nietzsche, so this is Mumford's call for an epiphany. He would
see an adequate sign; no sign, no "God," that he had understood was
adequate for the signified he fitfully tried to enunciate. The orthodoxy
of Mumford's assessment of the signs for God is evident. More than
three centuries earlier John Donne wrote, "We cannot take hold of God
as God. But as Father, Son, and Holy Ghost we can." His devotional
poems and many of his sermons present the problems of constancy that
beset the wayward human consciousness in the grip afforded by trinitarian signs. For Tate, ultimately the Roman Catholic, as for Donne,
the Anglo-Catholic, the trinitarian signs are privileged even if those in
communion through them are fallen. For Eiseley, the trinitarian signs
are honorable, perhaps even privileged with regard to the Son, but do
not afford adequate signification on the "religious pilgrimage" of his
life. All three understood words about God, *theologoi,* to be language
in process (lest it be dead, the dead letter, the mere static point that kill-
eth). Mumford, of that East Coast where America began as a western
nation, a land really a fringe between Europe and new-world future,

Mumford in American Adamic fashion proposes a new name for God, a sort of amorphous name in process for a God who is not yet.

The power of hatred and antilife, the power engaged for the mainte-nance of self against, in Mumford's terms, the emergence and creativity of others, is typically power applied to degrade the sign by censoring it or otherwise reducing its reference or by cynically manipulating the re-lationships of signifiers, signifieds, and referents: "So essential is lan-guage to man's humanness, so deep a source is it of his own creativity, that it is by no means an accident in our time that those who have tried to degrade man and enslave him have first debased and misused lan-guage, arbitrarily turning meanings inside out" (*CL,* 44). The point in more or less diluted form was made familiar, down to the level of intro-ductory college anthology, by George Orwell and lesser essayists even before the currency of poststructuralist meditations on the dynamic in-stability of the Saussurean bar between signifier and signified and on the instability of referentiality.

The enterprise in Mumford's project, we may feel, is two-sided. The more personal aspect reflects a manifest need to inveigh Amos-like at the multiple simulacra about him as a mid-twentieth-century American citizen. This side of his prophetic comportment is an existential chal-lenge to the old Western intellectual-ethical principle that a thing is not to be judged by its abuses. Was contemporary Christianity or even his-toric Christianity as conceived by his contemporaries too little charac-terized by its noble *essentia* to be acknowledged apart from its abuses? The more ideological and rhetorical aspect of Mumford's task involves drawing to attention the inadequacy of the available vocabularies for acknowledging the purposiveness and long-termness of life, an inade-quacy that he recognizes in others' language if not always in his own: "Among American theologians it has become the fashion to speak of ethics without religion as a mere cut flower, with no roots in the soil of life. . . . Just the opposite is the truth; for ethics lies in the common earth of life" (*CL,* 212). His image of the "soil of life" manifests the sense of persisting and constitutive history we noticed earlier in his im-age of a coral reef.

But *common earth* begs a question and reinforces the tenden-tiousness of Mumford's implicit definitions of both ethics and religion. What soil nourishes the ethics of the twentieth-century street criminal or corporate criminal or political tyrant and torturer? Where, on the other hand, did the historic religions come from if not from the soil of life? The holy men and holy women of the historic—as opposed to mythic— religions were not disembodied spirits. Mumford could, indeed, be accused of denying them any source but the soil of life, which he would not have been *quite* so swift as Eiseley to aver is itself mysterious. Yet, within a few pages Mumford—should we say in a different mood? certainly in a different vein—comes close to acknowledging as much, even as he parallels the Paul Tillich who wrote that a man's religion is that wherein he places his ultimate trust: "That which moves men to dramatic action in roles other than their natural ones is in fact their religion, no matter by what name they may call it. . . . That of the American pioneer was not Protestantism, but the conquest of nature and the winning of the frontier" (*CL*, 218–19). It has been averred that to a starving man, God may be a loaf of bread (though Gandhi fasted): ethics and religion might both appear to grow unavoidably *in* though not quite exclusively *out of* the natural roles and common soil. And growth can be healthy or malignant.

So what is afoot? To what dramatic action in what role is the Lewis Mumford moved who sat composing his book and himself in a decently provisioned family residence where the footfall of his dead son disquietingly did not sound? "Against the domination of the machine, we shall restore fresh energy to the word and the dream" (*CL*, 223).

The "we" battling the machine's hegemony are evidently prophetic criers: "If but one person in ten were fully *awakened* today . . ." (*CL*, 119; my emphasis). We cry prophetically from the margins, it seems: "More often than not, the excluded ones, the Gentiles, or the barbarians, the proletariat, the despised minority take the lead in . . . transformation" (*CL*, 217; Oswald Spengler, Arnold Toynbee, Henri Bergson, and Pitirim Sorokin variously disqualify—criers discordant or off-key). But we need not be a group; Mumford invokes James Clerk Maxwell's doctrine of singular points to affirm that at "moments of crisis . . . a single decisive personality, or a small group of informed and purposeful men," may reorient a society (*CL*, 228).

And the orientation, the direction and movement, that energetic word and dream may effect? The "dreams of love and brotherhood, the will to create a universal society of friends, alone hold promise of salvation" (*CL*, 252). Have such quintessentially American-utopian dreams any conclusive formulation in verbal precedent? Could they possibly have? Or that will to create? Is to suppose so not a Romantic dream of an omnific word, or at least a belated, print-era preacher's dream of a vocally compelling textual cry? (And to what extent is Mumford's society of friends a mythic displacement of lost family?)

Van Wyck Brooks seems to acknowledge no ontological or existential or semiotic problem more fundamental than the problem of *persuasion* attending his old friend's project: "His key word was always 'renewal.' . . . [He was] one of the few men who have not *ideas* but *an* idea, and he was to spend his life working this out." [40] That formulation nicely admits a focus within as well as without, a point of critical importance to which we must return: "Every suggestion put before [the reader] is meant as much for myself as for him" (*CL,* 252).

Mumford's goal of renewal may suggest that, dividedly, he wants an almost monolithic reintegration of the "new Babel of functions" that has confounded the "ancient Babel of tongues" (*CL,* 185). But he also knows there must be, and he seems to want, a syntax for the word of renewal. He proposes steps toward renewal for himself and others in the two dozen pages following. Observe what happens when his proposals, with their extensive commentary, are condensed to a list:

1. Hide yourself (the Epicurean injunction understood as a persisting, albeit partial, posture toward the world).
2. Withdraw daily for reflection.
3. Commit to fostering group life, so that "a new world will come into being."
4. Honor the "human scale . . . intimate cells."
5. Serve six months in a public service corps.
6. Reinforce parenting through communal action.
7. Give some love "to sea and river and soil."

(*CL,* 254–86)

So brutal a condensation invites an unfair comparison of Mumford's list with the one Jay Gatsby composed for himself when he was still young James Gatz—a list that included counsels like "Study needed inventions." It may also help explain the short-term political ineffectuality of the book, its ineffectuality in Van Wyck Brooks's terms. [41] Sea and river and soil were never so savagely violated before Mumford's book as

40. Van Wyck Brooks, *An Autobiography* (New York, 1965), 406.

41. Thomas S. W. Lewis (among others in a special issue) seems to complain that Mumford has not combated institutional ills more effectively ("Mumford and the Academy," *Salmagundi,* Summer, 1980, pp. 99–111).

since, although the countermovements he reinforced have clearly reduced the *increase* of carnage in major parts of the northern hemisphere. A roughly analogous estimate is possible about privacy and the effectuality of Mumford's defense of it. Parenting in the United States may be judged by the place children find in American life. That place is less psychologically hospitable or even healthy for many children than what the average child knows in any other developed culture or in many undeveloped cultures, Kampuchea and areas of famine apart.[42] The list abbreviated from Mumford's proposals highlights the spectacle of a man at least making and taking his own advice. His own efforts to foster group life were typically not so sharply focused as the work of his fellow American Academy member Tate in behalf of the republic of letters. But like someone planning a dinner party, he tried to put good deservers together—neighbor Americans with their neglected genius Herman Melville or with unfamiliar chapters of urban and technological history or with occluded but valid ideas for judging city and society. Similarly he honored the "human scale," the scale fostering intimacy, and warned against its negation, in *The Myth of the Machine.*

But a rhetoric of partial success, degree, and the like must not be allowed to obscure the conflictedness of Mumford's position. For one thing, he seems to have harbored something like pride in his complex conflictedness, making a point of revealing what could not have come as a total surprise to his old friend Brooks, who was working on a biography of him that never got published: "If you find yourself baffled in trying to put the parts together, be comforted by the report on my Rorschach test—the tester said that while he was usually able to summarize the results in a few hours, he spent two whole days trying to reconcile the contradictions in mine and make them fit a single personality!" (*VWB,* August 5, 1962). He would altogether replace the "authoritarian images" of Buddhism and Christianity on the grounds that they tend to be "ineffectual or tyrannous," at least "under our current nihilism" (*CL,* 118). Does that mean he looked toward replacing them with a new testament or covenant?[43] Rather more, with a new edict: "[One] must direct every habit and act and duty into a new channel: that which will

42. So Bruno Bettelheim, in a private conversation in 1983.

43. *Testament* is not a casual term here. Mumford wrote a fictionalized autobiographical piece, "The Little Testament of Bernard Martin" (1926) and contemplated a "Great Testament of James McMaster [*i.e.,* Patrick Geddes]" in 1935. See his *Works and Days* (New York, 1979), 118–19.

bring about unity and love" (*CL*, 119). Mumford's command is more estimable than that of Miss Havisham in *Great Expectations* ("Play!") but less compelling—and less compelling also than the biblical counsels of perfection they more nearly resemble.

It is hard to be a gospeler, a bearer of good news, if God is still under construction. Even Saint Augustine—no mealymouth, as Mumford knew, though he may not have known the relevant *De doctrina christiana*—modestly urges that when Scripture proves ambiguous, the preacher should choose the reading conducive to the reign of love (*ad regnem caritatis*). Still, if renewal is Mumford's word and gospel, as Brooks insists, he strives for his vision by mounting local attacks throughout *The Renewal of Life* on the hundred heresies traditionally said to attend every truth. But little attacks do not necessarily aggregate into any great victory; they may simply dissipate energy and attention. Few readers would conclude that in the first three volumes of *The Renewal of Life*, Mumford fritters away his resources; concerning the last volume, however, some might.

Mumford begins *The Conduct of Life* by writing, "For man existence is a continued process of self-fabrication and self-transcendence" (p. 5). He was aware that he began "this book a couple of months too early, for I have hardly yet had the time to get myself *clean* and *tidy* after finishing *Green Memories*" (*VWB*, July 12, 1947; my emphasis). As he continued working on the book, the scruffy prophet exhorting from the margin, he confided somberly to Brooks: "Parts of [*Green Memories*] are still unwritten, in a sense it will always be incomplete to me. . . . I have become a pariah except out of my own country" (*VWB*, November 13, 1947).

The movement from the first two volumes of *The Renewal of Life* to the latter two and especially to the last is from a largely metonymic discourse of desire—a bit of this to be praised or enjoyed, a bit of that to be deplored—to a more metaphoric discourse of love. That is, Mumford wrote increasingly as the would-be good son reconciled at last in adult fashion to Patrick Geddes, as the son who had adopted a supplemental father in Albert Schweitzer, and one might say, as the good grandson of the Ruskin who argued that there is no wealth but life. And he more clearly wrote in this fourth volume as the good father addressing a last letter of anxious paternal instruction and advice to Geddes Mumford and any foster brothers to Geddes still living. He wrote in affirmation of the whole man—the eighteenth-century man of light and the man of feeling, the post-Renaissance man of dream and postclassical citizenship, the twentieth-century man who is as likely a woman. He could use Emerson's title; he could not be Emerson. He could not prescribe or even identify or exemplify the "conduct of life" in any coolly Apollonian sense. But increasingly in *The Condition of Man* and *The Conduct of Life* he could exemplify *Americanus agonistes*, wrestling allegorically with adversarial metaphors and discourses and with their materializations in the institutions and more physical artifacts of culture. Uneasily, this discourse would be the good father to that in himself coming to terms with his marginality and mortality. Like Yeats's aging man who when young and lying had waved his leaves and flowers in the sun, now he could wither into the truth.

2

TO REDEEM THE TIME:
THE MYTH OF THE MACHINE

*Disorder increases with time because we measure time in
the direction in which disorder increases. This makes the
second law of thermodynamics almost trivial.*
>   —Stephen W. Hawking, *A Brief History of Time*

*. . . an Amos, a Hesiod, a Lao-tzu, deriding the cult of
power.*
>   —Lewis Mumford, *Technics and Human Development*

W hat occasion, what evil day, what question
prompted *The Myth of the Machine?* Although factors like occasion and identifiable
personal abilities cannot even in combination explain complex work of
human design with anything like predictive assurance or in anything
like definitive detail, the question admits of a sort of answer. The mass
killings driven by Axis paranoia and megalomania, the mass killings
aerially engineered by the Allies in a "logic" of war abhorrent to Mumford, the loss of one of only two children in the Mumford's nuclear family to infantry operations in what Mumford throughout deemed a just
and necessary war of defense against totalitarianism, thirty-five years of
study and writing about trends in Western technological and urban
life—these factors evidently at least fostered and enabled a conclusive,
panoramic vision of myth-driven, self-defeating behavior patterns
ramifying to crisis. Mumford had come to recognize that technological
myth tends to reinforce itself: man might be defined in terms of his origins, for example, as if origins were like the first gear in a gear train, and
his origins reductively defined as tool use; time might be fragmented
and reduced from history to separable events or event sequences. And

corresponding fabulations might not only be accepted from seductive purveyors, as Ernst Cassirer has argued,[1] but devoured, in Shakespeare's prophetic image, as if by rats that raven down their proper bane. Put another way: the reward for disentangling, above, the expressive conflicts in *The Condition of Man* and *The Conduct of Life* is to see in Mumford and his analyses of society the conflicts defined—with contrastive coherence—in the brilliant prophetic vision of the two volumes in *The Myth of the Machine: Technics and Human Development* and *The Pentagon of Power.*

"This whole book," he writes, "is a reasoned protest" against the "misleading habit" of "restricting the term 'invention' to mechanical appliances" (*T&HD,* 154). As he says in his prologue to the work, "Because I could find no clue to man's overwhelming commitment to his technology, even at the expense of his health, his physical safety, his mental balance, and his possible future development, . . . I was driven to re-examine the nature of man and the whole course of technological change" (*T&HD,* 10). Note the word *course,* fairly betokening his concern with dynamism and dynamics, to supplement what we might call taxonomics, and not least, his own dynamism: the work is "driven," and it is a "necessary move toward escaping the dire insufficiencies of current one-generation knowledge" (*T&HD,* 13).

He makes the compound move of reformulating and promulgating his megawarning in a conviction that the twentieth-century "enlargement of the Megamachine with increasing compulsiveness" is taking a form "increasingly coercive."[2] And he makes it as someone intensely implicated in the middle, looking both back and ahead, a hermeneutic

---

1. I am in the foregoing two sentences drawing on Ernst Cassirer's *The Myth of the State* (New Haven, 1946), *An Essay on Man* (New Haven, 1944), and *The Philosophy of Symbolic Forms* (New Haven, 1955–57). Volume I of the last work, *Language* (New Haven, 1955), is called "seminal" in the bibliography to *Technics and Human Development.* And I am indebted to John Michael Krois's fine analysis "Ernst Cassirer's Theory of Technology and Its Import for Social Philosophy," *Research in Philosophy and Technology,* V (1982), 209–22. See especially his Peircean remarks on "tychastic [mere chance] time" and disarticulated events. Compare Mumford: "History . . . could be discarded for practical purposes. Mechanical time was non-cumulative: it accompanied only changes of motion, changes of position, both open to external observation" (*CM,* 247).

2. "Technics and the Nature of Man," in *Philosophy and Technology: Readings in the Philosophical Problems of Technology,* ed. Carl Mitcham and Robert Mackey-(New York, 1983), 83.

historian who can recast the beginnings of cultural life, having seen the end of the "renewal of life," and vice versa.

Cassirer had perceived the increase in coercion by myth but had had difficulty viewing the monstrous myth of the state—differing from the myth of the machine in ways to be considered later—as other than a rationally conceived delusion foisted on the unwary. He offered no adequate explanation for the public's embrace of the fraud. Others, notably Jacques Ellul, have maintained that technology is augmenting itself in something like a mythic sense. But Mumford, as far as I am aware, has pioneered in arguing not merely for the antiquity of mythic thought (a commonplace) but also for the antiquity of the "modern" myth of the machine, and he has identified that myth and combated it more suggestively than anyone else.

He writes that in studying the pyramid age for *The City in History*, he came to recognize a Pharaonic megamachine animated by myth that had hitherto been "invisible to the archeologist and the historian" because "it was composed solely of human parts" (*T&HD*, 11–13, 188–96). This megamachine built the Great Pyramid at Giza in an astounding feat of controlled power and engineering precision. "No older vegetation myth, no fertility god, could establish this abstract order or detach so much power from immediate service to life" (*T&HD*, 198). In short—and Mumford's words do not explicitly go quite so far, quite so baldly—the megamachine that built the pyramids was possible in virtue of the abstraction of divine kingship and a recently augmented instrumentation of control. The Pharaoh's power was associated with astronomical order, something much more exclusively visual, remote, rigid, and abstractly mediated than, say, festive rituals with dance, and it worked with a new rigidity of prescription, control, and accountability through the instrumentality of writing, which had just become sufficiently acculturated for the purpose. "Exact reproduction of the messages and absolute compliance were both essential. . . . 'The scribe, he directeth every work that is in this land,' an Egyptian New Kingdom composition tells us" (*T&HD*, 192–94).

The first efflorescence of mythic technology, then, going far beyond mere tool use to megamechanistic pyramid building and militarism, is associated with the confluence of a totalizing view of authority and a potent new communicative resource.[3] So it has always been since.

3. Mumford's bibliographical annotations record that Loren Eiseley's "Fossil Man and Human Evolution" in the *Yearbook of Anthropology* for 1955 had "sparked" his

Mumford would insist on the parallel between "It must be as the god-king decrees" (where his decrees are in writing, allowing textually exact and replicable accountability) and "You can't stop progress." Mumford can also concede that some psychological and sociological satisfaction may result for the human parts of the megamachine, as for the users and usees of technology in general. Indeed, Mumford can resemble an Americanized José Ortega y Gasset in this: "The technical feat of escaping from the field of gravitation is trivial compared to man's escape from the brute unconsciousness of matter and the closed cycle of organic life" (*T&HD*, 35).[4] This has been able to transport men from the "daily grind" of agrarian village life, as he puts it in *The City in History*, to, at times, a titanic project. Compare the often-testified excitement of workers recruited from ordinary labs into the Manhattan Project, at Los Alamos, under J. Robert Oppenheimer.

The conclusion that the *escape* from "organic life" (how often and variously have we seen Mumford's *organic* as meliorative!) is ambivalent or even ominous because of its spatial and temporal abstractiveness represents not only a moral and political problem but in addition a problem that is hermeneutic. It is arguable that to understand the issue and to make it understandable is a prerequisite to deciding what should be done and to fixing on the means of doing it. But how is Mumford to convey a problem that has taken him a lifetime's studious experience to construe, a problem that he has construed as more ancient than most observers had imagined, one that concerns conditions both more reductive of the human temporal and experiential horizon and also more extensive than had been thought? A myth might indeed offer respite from "unbearable subjective confusion" (*T&HD*, 93; *cf.* "You can't argue with science"). But "no myth, however life-oriented, is wholly rational in its promptings" (*T&HD*, 149). Certainly not the myth of technological salvation.

---

approach; they also suggest the influence of Mircea Eliade, "especially" with regard to sky gods and sun worship.

4. See, most appositely, "Man the Technician," in José Ortega y Gasset's *History as a System, and Other Essays Toward a Philosophy of History*, trans. Helen Weyl *et al.* (New York, 1941), 87–161. This is acknowledged in a bibliographical note in *The Condition of Man* under the title of the 1941 edition, *Toward a Philosophy of History*, which Mumford cites in Chapter 7. In *Technics and Human Development*, he appreciatively acknowledges *The Revolt of the Masses* (New York, 1932) and *The Dehumanization of Art, and Other Writings on Art and Culture*, trans. Willard Trask (Princeton, 1948).

There are the usual hermeneutic difficulties of how there can be an understanding of the whole world without there being an understanding of the part, and vice versa, and of how there can be an understanding of the now without there being an understanding of the then or future, and vice versa. And there are the hermeneutic difficulties special to this problem, of conceiving a horizon historically large enough to frame ancient and modern phenomena meaningfully and existentially large enough to situate convincingly not only events, data, constructs, and institutional processes that have come to be omnipresent but also *ways of thought* that are dangerously seductive rationally and irrationally, not least in their presumption of totalizing inclusiveness. And the prophetic exposure of the myth of the machine and its divine pretensions will have the form not of an attack on any temple or on its money changers, nor of self-immolation, but of a printed text—a published, which is to say megamechanically produced, spatialized abstractive representation, instancing the very medium that has been historically intrinsic to the taxonomic, spatializing, atomizing abstraction of the mechanistic technologies and power myths under attack. Accordingly, it needs to be— who would seriously argue the point?—a long text, two volumes totaling some 730 pages of text and another 55 of annotated bibliographies. It needs, and has, pictures. It needs, and has, if not an assembly of voices expressing every point of view, at least a suggestive population that receive ostensibly direct quotation. It needs, and has—as it ranges through history since the Ice Age—very wide spectra of attention, from the abstract and the general to the concrete and the particular, with Mumford's attention to the particular evident not only in some of the pictures but also in his provision of emblematic instances and exemplary anecdotes. It needs, as Ortega and Cassirer helped Mumford know, the resolve to place man as technician in the context of essential man, and vice versa.

Mumford's hermeneutic project of understanding and representing on the turbulent boundary between myth and complex reality engenders for others the hermeneutic project of trying to situate his results in their realm of discourse and within their implicit horizon of shared concern, in some compromise between representation and brevity. Since *The Myth of the Machine* may not have the familiarity of more narrowly canonical literary works, it will help to lay out—in briefest form—the argument of its two volumes, *Technics and Human Development* and *The Pentagon of Power*.

The first quarter of the first volume's main text sets down the most general, and in part, the most conjectural, foundation of the chief argument. In a prologue and in chapters entitled "Mindfulness," "The Dreamtime Long Ago," and "The Gift of Tongues," Mumford elaborates with more sensitive concreteness than any of his ideological forerunners the view that "man is pre-eminently a mind-making, self-mastering, and self-designing animal" (*T&HD*, 9). When he turns to technics, by which he means all the equipment for life, and the production and use of it, he observes that "tool-technics" is "but a fragment of bio-technics" (*T&HD*, 7) and that biotechnics includes also language, an "infinitely more complex and sophisticated" culmination of world-making efforts than primitive tools. He objects to the misleading emphasis on—extending to misdefinition of—man as toolmaker on two other grounds as well: the preoccupation with tools, too narrowly defined, occludes *containers* (such as hearths, bins, canals, cities, and retorts; see also *The City in History*, Chapter 1), and obscures the positive contribution of technics—in his and Ortega's broad sense—to the "primal arts of order" like ritual and ceremony, by which, Mumford puts it as a suggestion, "man first confronted and overcame his own strangeness" (*T&HD*, 58–62). Chipping the given rock, ritually moving the given human body, voicing formulaic sounds (in effect making them quasi givens), reproducing conventional designs—all are ways of disciplining the extravagance and the potentially frightening waywardness of dream and fantasy. Technics can be a check on word magic. By the beginning of Chapter Five, then, Mumford has characterized men and women as self-inventors, both individually and collectively, in powerfully various ways ranging from the delusional and destructive to the imaginative and constructive, inevitably involved in a technics equally various in its propensity for ordering or for constricting and perverting human life.

For the remainder of *Technics and Human Development* and for all of *The Pentagon of Power*, Chapter Four, "The Gift of Tongues," is fundamental and critical. It grounds Mumford's sociopolitical and philosophical ideology in a conviction, of American and extended European ancestry, that man willfully *creates* his world, articulates it from givens and things made, by *action.* There is a deep-rooted belief in Western thinking that willful action is fundamental and that making and using language constitute the most consequential category of action.[5] But surely it was not only Otto Jespersen on linguistic origins and Johan Huizinga in *Homo Ludens* and other such texts but also Sophia Mumford, with infants Geddes and Alison, that prompted Mumford's sense of how the "semantics of concrete existence" (*T&HD,* 75) engenders from the "basic human dialogue" (*T&HD,* 85)—beginning with mother and infant—the shared and repeatable sign, the "true symbol," in a "marriage of internal need with external experience" (*T&HD,* 85).

Three chapters (Chapters Five through Seven) survey and meditate upon the surviving evidence of mesolithic and neolithic society. In part, Mumford celebrates the human achievement of those ten or dozen millennia as brilliant, since, in keeping with his "reasoned protest" (see above, p. 79), he does not take inventiveness to pertain only to mechanical appliances. He interprets the familiar hunter-gatherer patterns to be a dynamic progression from finding to making, as, for example, in the fashioning of artificial traps in imitation of natural ones. Close observation of animals and plants by hunters and gatherers encouraged an imitation of animal techniques and could lead to abstract thinking like classification into the edible, the edible if roasted, the fibrous, and so on. Moreover, it is arguable, Mumford thinks, that the "daily grind" and the attendant fabrication of *containers* for grains and other foodstuffs (the emphasis on containers is almost a Mumfordian trademark) showed the way to protective storage, enclosure, and even capital accumulation and social continuity (*T&HD,* 141).

In one respect, Mumford's interpretation is somewhat Heideggerian. Without quite speaking of an *essence* of technology, he envisages from the dawn of human society an increasingly influential technological *attitude* much like that which Heidegger identifies, an attitude that treats entities in the found world and their derivatives as *standing reserve,* that is, roughly, as mechanically controllable and as having being or worth solely as employed, especially for secondary operations.

Mumford mentions a concomitant, and somewhat contrasting, development during the earlier part of the period: a coming to predomi-

---

5. I am appreciatively agreeing, here, with the judgment of Don Ihde in the introduction to his fine *Existential Technics* (Albany, N.Y., 1983).

nance of feminine domestic skills in the pattern of life on farms with both horticulture and domesticated animals. On archaeological evidence, he associates the conjectured female dominance both with an "erotization of life" and with human sacrifice, the "dark shadow" of the "myth of maternity."[6] For Mumford, human sacrifice is not simply a scapegoating outburst triggered by a cultural crisis of desecrated distinctions, as René Girard sees it in *Violence and the Sacred*. Mumford thinks that, initially at least, it was conditioned by an association of blood with human and animal life in general, of blood with the onset of the menses and female reproductivity in particular, and—more speculatively—of the thinning of the tribe with the thinning of domestic plants. Perhaps, too, the lush growth above shallow graves played a part. But like Girard, and with a glance at Tiamat and Kali (the "dark side to woman's dominance"; *T&HD*, 148), he descries a relatively restricted practice of sacrifice that ramified in urban settings with the reemergence of masculine domination and became institutionalized as war. This he finds coinciding, far from accidentally, with the emergence of engineerlike specialists such as herdsmen, woodsmen, miners, and butchers.

But that is to anticipate the analysis of kingship and malign statism. Note that Mumford's suggestion seems to be that war, as a collective sacrifice, displaces individual and collective guilt onto the adversary, avoids the tedium of the feminized domestic daily grind, and materializes technological megalomania: the sequence seems to be to make bricks for a house, wall the city, build a kingdom, influence the elders, command the city, and then plunder the region and kill or enslave its people.

Before Mumford turns to the focus of his major ambivalence, the city, which is the synecdoche of institutionalized, literate human society,[7] he casts a long and reflective glance at the object of his lesser ambivalence, village life, in subsections entitled "The 'Neolithic' Syn-

6. Mumford seems not to have drawn the dismayingly circumstantial support he might have from Bruno Bettelheim's neglected but important *Symbolic Wounds: Puberty Rites and the Envious Male* (Glencoe, Ill., 1954), in which male vagina envy and creativity envy are seen as a profounder precursor of, among other aggressions, rationalizing and projective attibutions of penis envy ("We don't envy them; they envy us").

7. This disagrees with Morton White and Lucia White's Procrustean remarks on Mumford in their otherwise useful *The Intellectual Versus the City* (Cambridge, Mass., 1962).

thesis" and "Archaic Village Culture." Villages are significant places (none *too* specialized) for everyone, where one finds a *balance* (honorific word) among static and dynamic activities, and among outer and inner needs; nonetheless, after supposedly "adventurous experiments" to achieve domestication and seed cultivation, villages grew stultifying and remain so to this very day (*T&HD*, 159–62). Yet it was from the hamlet of Leedsville, near the village of Amenia, New York, that Mumford launched this critique, as well as several earlier ones, of urbanized, technologized man in action. The familiar, modestly renovated old house in its setting seems to have been an enabling margin—in the terms of literary genre, *both* georgic and pastoral. On the one hand there was real countryside and a real garden; on the other, the controlled simplification of pastoral that allowed an urbanite a place in a country landscape framed by selective connections with the more culturally conflicted world. Mumford had good mail service and telephones; road and rail networks connected him reliably with intellectual friends, with journalism, books, good libraries, and even collegiate lecture audiences (at CCNY, as the book's front matter acknowledges). But with the decades, whatever initially may have given a game-board quality to the place seems to have metamorphosed into essentiality.

The sketch of the healthy strength and—in his image from Sinclair Lewis—"Main Street" limitation of village life stands as a foil to most of the remainder of the book. Chapters Eight through Eleven are, in the main, a corrosive analysis of powermongering and its institutionalization, of psychological, political, and technological power. At the same time that he marshals data from primary and secondary texts and artifacts by familiar linear analytic procedures, along an axis of historical sequence, he presents a hermeneutic meditation on the consequences that the changes in lived experience had for the horizon of the third millennium B.C. In an otherwise puzzling series of subsections in these argumentatively crucial chapters, he proceeds not so much chronologically, not so much by strict progression from subordinate to superordinate parts or from cause to effect (considerations always present), as by a sort of hermeneutic dialectic: he seeks moments of knowing in order to believe, and of believing in order to know.

With utensils and vessels and bows and weaving, with, that is, enhanced materials management and power management, with improved tallying and delineating, with the communion and discipline of ritual in addition to the "ideated communication in language" (*T&HD*, 163) and the cooperative ordering brought by "taboo and rigorous custom," the dynamic complex was in being for a newly extensive social organization, hierarchical and expansive.[8] Mumford puts it that writing and associated abstractive, taxonomizing observation and thought (the tallying of produce, the planning of irrigation, the re-marking of boundaries after floods), along with a "shift of interest and authority from the gods of . . . fertility . . . to the gods of the sky" amplified the change of pattern and scale into a difference in kind: society became an invisible megamachine, the standing reserve for a technocratic god-king. The pyramids were built to exquisite exactitude without wheel, pulley, windlass, derrick, "or even any animal power except that of mechanized men" (*T&HD*, 166–68). The combination, as if magical, of the power of the hunter king's mace, the authority of the priest's astronomical knowledge and astrological mystifications, and the wealth of the god's harvest in the divine king's granary fomented megalopolitan and megalomanic projects.

What Mumford calls, in scare quotes, "civilization" he summarizes as a series of attributes of the cultural horizon within the "primal myth of divine kingship and the derivative myth of the machine." He mentions the centralization of political power, the separation of classes, the lifetime division of labor, the mechanization of production, the magnification of military power, the economic exploitation of the weak, slavery and forced labor (industrial and military), the invention and keeping of the written record, the growth of visual and musical arts, communication and economic intercourse for extended values and purposes, and a richer consciousness (*T&HD*, 186, 210). The written word especially belongs on the list, since it was the medium of *privileged* knowledge so long as it was largely a class monopoly. It was arguably the equivalent of today's higher mathematics joined to computer technology (*T&HD*, 199, 273, 285) or of the quasi-controlling pseudoinformation of Mumford's bureaucratized "communication-machines" (*T&HD*, 201).[9]

8. See *T&HD*, 114–15; whether the bow, the earliest of complex machines, was first used to propel an arrow, to drive a bit (to make a hole or a fire), or to make musical notes, pizzicati, is a question Mumford may have had prompted by Ortega's essay on technology or may simply have thought of.

9. This is a point made earlier by Allen Tate in "The Man of Letters in the Modern World" (*Hudson Review*, V [1952], 335–45; reprinted in *CE*, 379–93). Mumford cer-

Four of the attributes at the end of the list—records, arts, extended creativity, and richer consciousness (see also *T&HD*, 208) represent in brief what Mumford treats more expansively as the rewards flowing directly enough from the megamachine to forestall some of the rebelliousness against it. War appears as a logically subordinate but existentially enormous dark subset among these compensations. He defines it as *not* a genetic disposition but a "cultural institution" (*T&HD*, 216) and argues that, from the first, it has provided a seductive mix, even an addictive mix, of compulsive control and hysteric expressivity: a "compulsive collective pattern of orderliness" in army discipline counters the anxiety of circumstance; in battle and plunder there are compensatory releases from the "daily grind"; there is mock creativity or, as I would prefer to call it, stolen significance, another kind of relief from the megamachine's effect of subdividing labor below the critical mass needed for it to have social significance and to give existential satisfaction. Milton's Satan surveys his forces and exults; George C. Scott, in the movie *Patton*, surveys his "cumbersome luggage of war" with just a hint of saving suspicion that he exults too much.

Short of war's hysteria but perhaps framing it, and certainly more pervasive, is the human destruction attendant on what Mumford describes as the megamachine's degradation of work (*T&HD*, 212, 242). The sense that all work is degrading to the human spirit has spread insensibly from the megamachine to every manual occupation. It will be necessary to return to the matter of work blessed and cursed and the question of whether the megamachine's bad features, as Mumford lists them, imply a specific program of resistance for a great society, either in fact or in Mumford's judgment (see below, pp. 96–106). Relevant in that connection will be his acknowledgment of the "dream of a Golden Age, part memory, part myth" (*T&HD*, 240). But suffice it here to note a long transition, covered through many pages of the final two chapters of *Technics and Human Development*—chapters entitled "Invention and the Arts" and "Pioneers in Mechanization"—in which during the Middle Ages the ancient megamachines with almost exclusively human parts and avowed religious sanction dissolved, to return in modern recrudescences with enormous mechanical reinforcement and more subtly

---

tainly was referring to propagandistic manipulations of history, of current events, and of consumer behavior by advertising. For a brilliant postscript illustration, see Jonathan Schell, *The Time of Illusion* (New York, 1976).

encoded religious sanction. Mumford canvasses the rise of what he, like Karl Jaspers, elects to call the axial religions, as he also examines the breakup of the Roman Empire and the ongoing development of technics, especially its static developments and its extrapolations from the neolithic art of containers (*T&HD*, 250). In the four volumes of *The Renewal of Life* and in *The City in History,* enduring cities such as Rome, rising medieval cities, medieval technics, and religious developments all have turns in the foreground. Here they enter more briefly, as ambiguous counters to the megamechanic attenuation of life and to the ultimate reformulation of the megamachine in modern form. Mumford finds in the modern megamachine, on the one hand, an aggrandizement of the individual soul, a mechanized deliverance from some drudgery of sawing and grinding and pounding, a number of small-power aids for familial or quasi-familial shops, the "Benedictine blessing" of ordered and productive communal monastic life in refuge from widespread want and disorder, as well as the benefits that arose from the early guilds, and on the other hand, an acclimatization to pervasive mechanism as well as dangerous temptations to unbridled enlargement and extension in mechanical, ecclesiastical, and political power—the narcotic fruits of success.

The Benedictine rule and the church-tower clock temporalized the day in ways more abstractly taxonomic than any before, notwithstanding that the divisions could be for worship or society. The guilds could become restrictive, material gain could be fetishized, and the church could become secular and megalomanic. "The monk, the soldier, the merchant, and the new natural philosophers" were each, like Ibsen's John Gabriel Borkman, "ready to forgo love and to sacrifice life, in order to exercise power" (*T&HD*, 277), especially through the cash nexus (*T&HD*, 281). In the stalemate of the Christian sects, Francis Bacon's *New Atlantis* might be taken to speak for a gratifyingly material and secularly ecumenical "alternative way of reaching Heaven" (*T&HD*, 283). And between "the twelfth and the sixteenth" centuries, a "new technical assemblage" came into being that enabled a "world revolution, strangely similar, in all its main assumptions and goals, to those of the Pyramid Age" but relying on the assemblage of water- and windmill, mechanical clock, printing press, and optical lens (*T&HD*, 287).

Physics, as we know it, will be over in six months.

—Max Born, 1928, quoted by Stephen Hawking
in *A Brief History of Time*

The Pentagon of Power, Volume Two of *The Myth of the Machine,* devotes itself to an analytic history of the resurrected megamachine, to its ramifications and implications, and considerably less emphatically to countermovements and alternatives. The discourse is very markedly one of hermeneutic systole and diastole. In its variation, repetition, and reprise, Mumford presumably seeks to know ends by knowing beginnings, to know beginnings by knowing ends, to know the whole by knowing parts, and to know the parts by knowing the whole, as he addresses the question of why we seem so overwhelmingly committed to technology.

He begins by arguing that the age of exploration was significantly bimodal (I follow the suggestiveness only a step or two here): it was astral and abstract by way of mathematics and the new telescope, and concrete yet phantasmal by way of geographic searches for new Edens of endless innocence or of endless power and plunder (the latter point Eiseley's too). Oceanic navigational practices connected the two modes of the age, and medieval theology, the queen of the sciences, provided traditions and conventions for dealing in abstractions, albeit nonmathematical ones. The debt abstract thinking owes theology was slighted by Enlightenment historians because of anticlerical bias, according to Mumford the corrector (*TPOP,* 6).

Ironically, the Spaniards found in Mexico and Peru the "same institutional complex that had shackled civilization since its beginnings in Mesopotamia and Egypt: slavery, caste, war, divine kingship, and even the religious sacrifice of human victims" (*TPOP,* 9), not to forget grandiose state building projects. Sumerian, Mayan, and Indic cultures, Mumford notes in one of his most far-reaching insights, had "coupled human destiny with long vistas of abstract calendar time" but had not known much of the "cumulative results of history" nor articulated them as "cultural achievements that marked the successive generations." Yet, he says, the often-touted Renaissance discovery of history was, in its fantasies of a New World, typically an "attempt to escape time and . . . tradition and history . . . by changing it for unoccupied space" (*TPOP,* 13–14)—or for space occupied by merely a few "natives," a standing reserve to be erased or worked to death.

He can suggest that "for the first time . . . the future . . was more attractive than the past." He does not quite say that this was because of the spreading technological attitude that made the past a standing reserve, a parts depot to supply a standing-reserve future that was conceived either as a mechanical extrapolation from the present or in an utterly detached fantasy. Ben Jonson satirized both attitudes in *The Alchemist*, the former in the pedestrian Abel Drugger, who wanted the "philosopher's stone" to give him an extra margin of success for his modish tobacco shop, the latter in the febrile imaginings of Sir Epicure Mammon, who sought infantile omnipotence, with sex.

As evidence for the pervasiveness of the myth of the machine, Mumford adverts to the preponderance of credit science as currently constituted gives to the abstract and mathematical, to the serious neglect of "the concrete, the empirical," in its prizing of *facta* over *data*.[10] Concrete procedures, observations, and logbook entries contributed as much to scientific method and the scientific world view as abstract "capitalist accounting" did, according to Mumford (*TPOP*, 18–21).

Indeed, the very form of discourse throughout the two volumes of *The Myth of the Machine* is a kind of ostensive definition and polemical demonstration of concrete thinking, in opposition to one-sided abstraction: it has argument, witness, emblematic example, anecdote, pictorial illustration in the same way his *Technics and Civilization* did almost two generations earlier—though that book was less rich in its knowledge or counterpointing. Mumford is not alone in this corrective move: he cites Lynn Thorndike, and might have cited G. Evelyn Hutchinson on medieval naturalistic carvings. But other critics of technology, such as Heidegger, Ellul, and the Frankfurt school, have been accused of excessive abstraction, and each reader may wish to conduct an internal audit on the matter.

The next four chapters, to speak broadly, alternate in their predominant focus from politics to science and technology, first for the sixteenth century and then for the seventeenth. But for the moment, I want to scant the world of infant science, with its so-called, so wrongly called, primary qualities, the world of quantifiable space, time, mass, motion, and gravitation. What troubled me as most questionable about Mumford's argument when many of these pages appeared in the *New Yorker*

---

10. James Gleick has fascinatingly confirmed the charge and documented the post-1960s countermovement in his *Chaos: Making a New Science* (New York, 1987).

in October, 1970, was the conscious overtness he ascribed to the "return of the sun god," in the words of his chapter title. I shall present as schematically as I can the argumentative skeleton of Mumford's subthesis about that.

Into the context of factors friendly to mechanization and quantified competition and power struggle—factors exhibited in Chapters Ten to Twelve of *Technics and Human Development* and in Chapter One of *The Pentagon of Power*—there fell in the 1540s Copernicus' book on the revolution of celestial bodies, Vesalius' book on human anatomy (both in 1543), Cardan's book on algebra (in 1545), and Fracastoro's book on the germ theory of disease (in 1546). As books, these were, unlike voices, manipulable objects within space, and each was about objects or relationships in space that are manipulable or at least measurable. The immediate effect of Copernican heliocentrism was not, Mumford provocatively argues, so much to demote man as to aggrandize the sun and science (*TPOP*, 28–30). The sun became "no longer a satellite or servant, but the master of human existence" in terms of gravitational, heating, and illuminative power. And the ruling class "knew" what had happened, partly because of their patronage of the new keepers of the mysteries of solar and celestial mechanics. Moreover, in a hierarchical age with relentlessly analogizing, emblematizing habits of mind, it was predictable although not inevitable that the monarch would be profoundly linked with the sun and its godlike power. One recalls Shakespeare's Richard the Third positing a winter of discontent made glorious spring by a son of York, and Spenser's "Colin Clouts" on the "Sun's life giving light." One recalls, too, the sunburst image of Gloriana, Queen Elizabeth the First, who did not need a confidence seated in the pun on *son* and *sun* in the English language to feel herself the Lord's anointed and vicegerent, any more than did Louis XIV in seventeenth-century France. Mumford observes that the new myth of centralized power which was tied to solar mechanics slighted feminine generativity, in the manner of the Egyptian myth of Atum-Re, who, self-created, created the universe and the other gods by masturbation. Elizabeth bore no children, though urged to marry and procreate.

In 1582, the pope revised the calendar, extending the confusion of astronomical mechanics with politics. Even astrology, Mumford adds, catered to, without ever satisfying, the lust for deterministic control: "Exploration, invention, conquest, colonization all centered on immediate fulfillment. Now, not the hereafter, was what counted" (*TPOP*, 34).

The now that counted means not the instantaneously immediate, of course, but the proximate future articulated by the designs of a *roi soleil* as surely as the heavenly sun makes the seeds germinate. Mumford implies a cult of the soon, equivalent to what Eiseley calls anticipatory man. Although Mumford does not believe that technological or even technocratic interests *necessarily* articulate a society in tychastic time, he does believe that historic time, if richly textualized, *might* be assimilated to release man "from the grip of his unconscious past" (*TPOP*, 39). And he does interpret the governing preoccupations of the Renaissance as being increasingly with mass and motion, *as if* they were in cosmic isolation from earthly ecology, from the subjectivity of the earthly observer, and from implications in history: "Abstract motion took possession of the Western mind" (*TPOP*, 33, 37, 45−46).[11]

Mumford cites as recent support for his interpretation E. M. W. Tillyard's pioneering *Elizabethan World Picture* (1944), which his bibliographical entry generously calls "brief but penetrating," and Mircea Eliade's much more wide-ranging *Patterns in Comparative Religion* (1958; annotated in *T&HD* for its chapters "on Sky Gods and Sun-Worship"). He quotes Johannes Kepler, who in 1603 wrote on the sun as the "most excellent . . . than which there is no greater star; which single and alone is the producer, conserver, and warmer of all things: it is a fountain of light, rich in fruitful heat . . . and which alone we should judge worthy of the Most High God, *should he be pleased with a material domicile*" (*TPOP*, 35; my emphasis). So much for the orthodox doctrine of the incarnation! One sees more of what the Sun King was claiming. Mumford does not seem to have known Michel Foucault's *Les Mots et les Choses* (1966), which was somewhat likelier to have come to his attention, with its fine second chapter on Renaissance analogizing, when it appeared in America translated as *The Order of Things* (1971). His own post-Enlightenment anticlericalism may have repressed his recollections of *Paradise Lost*, which he seems to have read in the thirties.

Recall Book IV of *Paradise Lost* and the relatively neglected Book VIII. Milton once uses the image of the "fountain of light" for God, but as if to answer Kepler and all Keplerians, has the voyaging colonizer Satan stop and admire the sun almost worshipfully. Satan also frowns in rancorous envy of the newly created world. Immediately after the pro-

---

11. Mumford's formulation of this may owe something to Alfred North Whitehead's celebrated *Science and the Modern World* (New York, 1925).

digious story of that creation, in Book VII, Adam, at the beginning of Book VIII, frowns and questions, indeed challenges, the idea that the celestial economy of the glorious sun and stars has nothing to do but afford light to this point, this speck, this earth. The archangel Raphael rebukes him on two grounds: for confusing size and brilliance with *goodness,* and, drawing a stark contrast to the fecundity just represented in the creation story in Book VII, for overlooking the *barrenness* of the sun itself, however much its warmth might be a divinely created agent of fertility. It was Milton, arguably the acutest commentator on his own age, who prompted me to consider that perhaps Mumford had read the record right.

Mumford's argument for the renascence of the sun god aside, his more familiar argument—that astronomical *kinds* of knowledge, of force, mass, and distance, as if in a human vacuum, dominated the "mechanized world picture" (Chapter Three) and "science as technology" (Chapter Five)—is introduced by an account of "Kepler's dream," in the *Somnium,* of 1635. He offers this as an emblem of the audacity that marked the beginning of a new technological era, and equally as an emblem of the "hot subjective pressures" that drove seventeenth-century political and technical doings and darings (*TPOP,* 46–47). The remarkably realistic voyage Kepler envisioned to a nightmare moon elicits a characteristic Mumfordian query, a "serious question that is impossible to answer and fruitless to speculate upon: *Why did Kepler suppose that a journey to such a planet was worth the effort?*" Mumford, whispering that he would ne'er speculate, speculated: "Why did the utmost achievements of technology . . . terminate in fantasies of shapeless monsters and cruel deaths, such as often haunt the cribs of little children? If we had an answer to this question, many other manifestations of the life-negating irrationalities that now threaten man's very survival would perhaps be sufficiently intelligible to overcome" (*TPOP,* 48–49). Could he have speculated further about, or at least focused here, the fantasies of power and control he attributes at times to ancient and modern despots? Children of any age or era may feel compelled, in the current phrase, to "test the limits" if they have not internalized limits or have not felt caring limits and if they feel the kind of self-doubt that compels self-assertion. Are we not inclined to believe that irrational, destructive, and megalomanic acts of self-assertion are animated—like many childhood nightmares—by separation anxieties, indeed, by fear of nullity? Thom Gunn's bikers of "On the Move" accelerate the motocycles of the "created will" in order not to be nowhere. As with the "Nowhere Man" of *Yellow Submarine,* being on the way to no place in particular is preferable to facing the frightening nowhere of repose.

The formulation suggests a lack, even an absence, in the latter-day world view. Mumford knew Huizinga's *Waning of the Middle Ages* and its definition of decadence as essentially a matter of something left out or lost. In a partly retrospective Chapter Six he mourns the medieval polytechnics that was lost when it was transformed by political absolutism and capitalist enterprise. The transformation that produced the industrial revolution, conventionally defined, of the eighteenth century was from labor loving to labor saving, he believes (*TPOP,* 136–37). In it the privileging of political absolutism and capitalist enterprise (*TPOP,* 131) was, then, a decadence, by Huizinga's definition: lovable work has been lost or degraded. Indeed, what Mumford in Chapter Seven, on "mass production and human automation" (and *passim*) calls the pentagon of power is a kind of core of values that is hollow with regard to love. He names the component aspirations of what he also calls the "power complex" as *power, productivity, profit,* and *publicity.* And he argues that in the modern era those aspirations have fused into a "subjective unity," a sort of bad center with a compelling quality analogous to that of the "pleasure center" in the brain (*TPOP,* 157, 167). He has a telling exhibit in substantiation of his thesis in the usage of the term *progress*—which names a fifth aspiration—in countless ways that center upon "abstractions of time and space and motion," like the higher, farther, faster—how Olympian!—what takes less time, and so on (*TPOP,* 24). Moroever, as he argues at length, on into a critical review of eighteenth- and nineteenth-century utopian literature (Chapter Eight, "Progress as 'Science Fiction'"), "all automatic systems are closed and limited" (*TPOP,* 191) whereas nature is prodigally not.[12]

The complex openness of natural organic life—the counterentropic tendencies of natural growth, development, and ramifying complexity (*e.g., TPOP,* 202), and the dynamic balancing of tendencies toward continuity against proliferating variation (*e.g., TPOP,* 209)—is the countertext of Chapter Eight. Mumford thus glosses utopian fictions from Thomas More to Edward Bellamy as reductively unnatural—as

12. For poignant accounts of recent paradigm-challenging scientists trying to take account of that general fact rather than ignore it, see Gleick, *Chaos.*

variations on a masked theme of iron automation, changing later to the theme of an iron automation of expertise, overtly and satirically so in Aldous Huxley's dystopian *Brave New World*.[13]

Mumford attempts to distinguish between causes and anticipations. Literary and scientific utopias had no serious effect on technics or on the "process of mechanization." They were not causal influences. Yet they were, and remain, he thinks, useful imaginative "cross-sections" of the all too imminent "collective termitary we have been bringing into existence" (*TPOP*, 211–12). His real complaint with utopian writers is that they have merely extrapolated, constructing at most a "de-corticated" *reductio ad absurdum* of a brave new world (*TPOP*, 227) instead of an imaginative transformation. Consider:

> In turning his back on the realities of subjective life, Descartes rejected the possibility of creating a unified world picture that would do justice to every aspect of human experience—that indispensable precondition for the "next development of man." (*TPOP*, 94)

> A sound and viable technology . . . must, rather, seek as in an organic system, to provide the right quantity of the right quality at the right time and the right place in the right order for the right purpose. To this end, deliberate regulation and direction, in order to ensure continued growth and creativity of the human personalities . . . must govern our plans. (*TPOP*, 128)

> To build up human autonomy, to control quantitative expansion, to encourage creativity, and, above all, to overcome and finally eliminate the original traumas that accompanied the rise of civilization—of these fundamental needs there is no utopian hint. (*TPOP*, 219)

Indispensable, must, ensure, finally eliminate—the language of the most radical utopian of all, the American prophetic utopian.[14]

The somewhat heterogeneous follow-up chapter, with the punning

13. What had been gradual "disenchantment with the utopian tradition" over fifty years had become "now complete," Lasch argues, citing "Progress as 'Science Fiction'" ("Mumford and the Myth of the Machine," *Salmagundi*, Summer, 1980, pp. 25–27). Lash is surely right about Mumford and traditional utopian texts, perhaps right that Mumford acknowledges the "radical imperfectibility of human nature," but mistaken or neglectful about Mumford's ultra-American hope for progress transcending science fiction.

14. I come to this recognition by sharing in the insights of my colleague James Hardy, who contrasts American prophetic utopianism with European adjusting, as represented by such an American socialist as Michael Harrington and by Thomas More's monastic Utopia.

title "The Nucleation of Power," adjudges Henry Adams the prescient "best" interpreter of the changing relation of technology and the cultural situation. Mumford honors him as a dismayingly unheeded prophetic forebear despite Adams' "inexact knowledge," "dubious mathematics," preoccupation with energy as the controlling single factor, and Calvinistic determinism. Mumford discerns in Adams what was mainly a version of himself, but a self he might have known less well without the mirror of Adams.[15] He saw incipiently a lifesaving new kind of historian: "By his stress on the potential future, he had moved from the world of serial time and causality . . . to the organic order of temporal duration, of phylogeny and social inheritance (memory and history) in which both past and future intersect in the present. In this organic world purpose superimposes itself on process and partly transforms it" (TPOP, 235).

Hitlerism stands in Mumford's view as a more dramatic illustration of Adams' prescience than Stalinism. That is because, although both tended—in Albert Speer's phrase—to mechanize the lower leadership, in nazism the "most virulent forms of racialism, existentialism, blood-and-soilism" were attached "cunningly to reputable sentiments and genial emotional needs that had been left out of the mechanical world picture" (TPOP, 249).

Chapters Ten and Twelve, "The New Megamachine" and "The Megatechnic Wasteland," may be construed as a closely related series of ten subsectional meditations on features and concomitants of the new, self-amplifying pentagon of power. The secret of information power, he argues, is secrecy itself. And he notes provocatively what is at once a continuation and a change: in the pyramid age, writing and the command of writing were a secret code monopolized by the most powerful; now the equivalent mark of power is secrecy by fiat or by encodement, including encodement in arcane specialization. Mumford might be suspected of verging on paranoia in styling the congeries of the AEC, FBI, CIA, and NSA an equivalent of the Egyptian dynastic establishment, immune to public criticism. Since the death of the self-dynastic and ulti-

15. That Adams would have had difficulty recognizing the characterization, I conclude mainly from Joseph Kronick's persuasive delineation of an "intellectual horticulturist," desperate with the loss of history, who saw that science is metaphysical but who both noted and exemplified the dangerous human predilection for myth (*American Poetics of History: From Emerson to the Moderns* [Baton Rouge, 1984], Chapter 4, "Henry Adams and the Philosophy of History").

mately vile J. Edgar Hoover, the FBI has come in for some public criticism. The CIA has borne even more, and the AEC has met increasing challenge. But the NSA remains a very cryptic organization indeed, and a concerned citizen may wonder doubtfully whether the private effects that public criticisms have had on any of those organizations have served the public weal. Hoover lived in office until his death of natural causes. Is it paradoxical that much of NSA's activity is precisely in privileged, secret interception, decoding and unveiling of other people's secret communications?

Scientists have too often become interested parties to power projects rather than disinterested researchers, Mumford complains, and computer power threatens to create an "all-seeing eye" over the information about anyone's life which might enter disparate computer memories. Meanwhile, in the populous echelons of middle management, "the model for Organization Man is the machine itself. . . . Like Himmler, he may even be a 'good family man'" (TPOP, 278–79). Some such notion of the organization man had become almost a cliché by the year Mumford wrote.[16] But he invigorates the cliché with his anatomy of the weaknesses of computer power and his observation that computer power is hobbled by the limitations of its input and that the input is distorted by the ignorance and group-thinkery of its ministrants and is at very best reductive of the richness and the flow of human life in nature (TPOP, 273). In addition, he believes the desire that drives ministrants to the technopower complex to be wretchedly meager: "The final demand of Organization Man . . . [is] to make the world over in his own shrunken image" (TPOP, 281).

Mumford's debunking of Marshall McLuhan's dated notion of the "global village" and Pierre Teilhard de Chardin's of the "noosphere" may be passed by in favor of his more far-reaching critique of "space," in the NASA sense: "The new dynamic forms of the pyramid-complex—the skyscrapers, the atomic reactors, the nuclear weapons, the superhighways, the space rockets, the underground control centers"— all betoken, he asserts, a sort of neo-Calvinist elite whose power over the manipulated mass of men is a quasi salvation (TPOP, 300–301). Though well read in Freud, Mumford minimizes his attention to the phallic significance in space rocketry in order to consider the interest in

16. Mumford characterizes William H. Whyte's The Organization Man (Garden City, N.Y., 1956), as "shrewd observations, though . . . limited."

space in terms of the related ideas of control trip and power trip. His more particular assertion of an equivalence between the Pharaonic burial cell at the heart of a pyramid (with "miniaturized equipment" for "travel to Heaven") and the manned space capsule may be noted as disturbingly suggestive, even if less conclusive than he believes. Certainly he is correct—and this before the obscene misfeasance of the *Challenger* disaster—that space capsules and their destinations differ radically from any place where human life and mind have flourished in the scores of millennia before. And when he was writing, space research and development were preempting "funds and personnel from every secular activity" (*TPOP*, 309), as the so assertively called Strategic Defense Initiative has threatened to do throughout the eighties and nineties.

Two other arguments, enunciated but not developed in the chapter or the volume, seem to me equally far-reaching. One is that the megamachine or power-worshiping complex suffers from an *overparticularization of purpose* along with its arrogantly *overgeneral claims* to the adequacy of its knowledge and method. Projects like SDI or the interstate highway system or "putting a man on the moon in this decade"—or on Mars in this century—need to be weighed against larger alternatives and larger purposes, he implies, than the priests of primary-qualities power are in the habit of invoking. Although Mumford quotes the phrase "things such as make most for the uses of life" from Bacon's *Advancement of Learning* in the epigraph to his own epilogue, he is more ambivalent than Eiseley about its larger Baconian context. What he actually wants, without quite being willing to put it so, is an Augustinian test that derives from *De doctrina christiana:* a test that evaluates a cultural option, as it would a scriptural interpretation, by whether it conduces to the reign of love (*ad regnem caritatis*). Compare the self-interested lobbyists for light-water nuclear power reactors.

The other adumbrated argument of his that seems to me far-reaching is that the megamachine's "increasingly coherent structure, self-organizing and self-expanding," in "both its purposefulness and its highly complex ultimate character . . . resembles language" (*TPOP*, 311–12). Neither Mumford nor anyone else has worked out in detail the implications of that hypothesis with regard to megamechanic syntaxes of desire, but *The Myth of the Machine* as a whole affirms that neurotic powermongering and control fetishism are the driving forces of the megamachine. No Hayden White or Paul Ricoeur has arisen to pro-

vide a "tropics" with regard to the megamachine or a temporal metaphorics of power-complex discourse. But Mumford's long prophetic history and argument are the most suggestive start we have toward a
neorhetorical exegesis of such metaphoric likenesses as those with the
pyramid age, of metonymic fractionations and reductions of powermongering, of synecdochic instances, and above all, of the ironic dis-
proportion between signs of the power complex and the horizon of nature and human life that is their ostensible reference.

His own most fundamental asseverations, in their minimally metaphoric and broadly general language, are, I think, actually synecdochic,
out of a passionately lived family life uniquely yet typically fulfilled:
"Life begins, even in the lowest organisms, in physical accretion and ecological association, and develops in the highest organisms into mutual
support, loving reproduction, and hopeful renewal" (TPOP, 317). It
hardly matters that he is controverting Teilhard de Chardin, as he was
earlier the "ingrained fatalism" of Ellul (TPOP, 290–91): the essential
point is that the prophet paints a profoundly dark picture and paints it
with convincing fidelity, in part because of the very brightness of the
frame,[17] and in part because of his communal allusiveness.

Somewhat as Henry Adams was the prophetic forebear to whom
Mumford genuflected in Chapter Eleven, Alexis de Tocqueville receives
a bow in Chapter Twelve, "Promises, Bribes, Threats," because of his
prescience about the potentially inordinate costs in mass-producing con-
sumables (TPOP, 344–45). Saint Augustine remains an unacknowl-
edged godfather in the distinction between a power-complex society of
mass production in which "continence and selectivity" may be "punish-
able vices" and a more sensitive society for which a framing criterion is
that "all these goods remain valuable only if more important human
concerns are not overlooked or eradicated" (TPOP, 332–33). In a similar Augustinian vein, a subsection, "Quantification Without Qualifica-
tion," concerns the distinction between power, as in assembly-line production, and love, as in craft-shop production.

Of course, it is possible to feel that the argument is exaggerated here,
that the cybernetic revolution offers deliverance, by automation, from
boring production-line work even while facilitating great variation in

17. My second thought on recognizing this was that the trope precisely reverses one
Louis Martz used in his lectures many years ago, about *Paradise Lost:* "a bright picture
with a dark frame."

computer-programmed production. And has there not in the past generation been something of a revival of handicrafts? But even apart from the stolen profits accruing to firms that automate and throw on public assistance their unemployed workers, one must also think of the obsolescence built into automobiles, of the tobacco production subsidized by the government and the tobacco advertising not outlawed, of the World Trade Center rising like an exhalation in overcrowded Manhattan. One may conclude that Mumford's argument in general and particular has been profoundly synecdochic: "Scarcity remains: admittedly not of machine-fabricated material goods or of mechanical services, but of anything that suggests the possibility of a richer personal development based upon other values than productivity, speed, power, prestige, pecuniary profit" (*TPOP,* 337).

Chapter Thirteen attends to instabilities in and impediments to the megamachine, as its title, "Demoralization and Insurgence," straightforwardly names them. More ambiguous and ambivalent is Mumford's actual treatment of those forces. On the one hand, the rigidity and reductiveness of the megamachine, its innate unresponsiveness to hostile feedback, make it vulnerable; reality is even more arrogant than myth, as an ecclesiastical friend of mine used to remark. On the other hand, the megamachine's power and seductive success have attenuated such "safeguards" to cultural stability as rural life, village life, religion, and—only a bit paradoxically—migration. Mumford the hyperbolist describes the safeguards as "vanished" (*TPOP,* 350–52). Certainly the inadequacy of supporting structures and nurturance has left countless young with what has conventionally become termed an identity crisis.[18] And even more ominous are the crime statistics, the overlapping, darker manifestation of regression and "anti-life." Street crime is the mechanistic objectifying of victims by victimizers, who apparently have often mechanized themselves with drugs.[19]

More problematically, Mumford, who in this context as in others sees fantasy as a blessing always in danger of deranging man, here insists that a childlike or "primitive" need for and "gift for enjoying exact

18. Mumford in text and bibliography praises Erik Erikson, the promulgator of the term, but does not list his fine *Identity, Youth, and Crisis* (New York, 1968) or *Childhood and Society* (New York, 1950).

19. Philip Slater, in his fine essay *The Pursuit of Loneliness: American Culture at the Breaking Point* (Boston, 1970), first moved me to consider the mechanistic implications of "turning on."

repetition lies at the bottom of human culture," and he complains that the power complex "has transferred all the stabilizing repetitive process from the organism to the machine" (*TPOP*, 369–70). True, print culture—an element of the power complex—has largely superseded oral epic formulas, proverbs, and other verbal mnemonics that work, markedly, by repetition. But nursery rhymes have not disappeared, and the use of any language entails endless repetition of largely conventional syntactic patterns, such as subject-verb-object in English. Hence the repetition of language or bodily movements so obviously dear to children, like the rejection of threatening patterns in the "verbal salad" of schizophrenics and the repetitiousness of hysterics, is an effort at control.[20] Control, so obviously necessary for survival, evidently becomes like the water bucket–carrying broom that the sorcerer's apprentice mobilized, an engulfing problem when out of control by reason of fetishism or any other disorder of desire. Such is Mumford's reading of the influence on rising generations of our "over-heated technology": "Basic conditions for mental stability—accepted criteria of values, accepted norms of conduct, recognizable faces, buildings, landmarks, recurrent vocational duties and rituals—are constantly being undermined; and as a result our whole power-driven civilization is turning into a blank page, torn to shreds from within by psychotic violence" (*TPOP*, 370–71). The "blank page" with a nice economy genuflects toward Carl Jung and excoriates John Locke.

In Chapter Fourteen, Mumford leads a tour of the intellectual ports of call of a better than Baconian New Organum. The tutelary genius is Darwin, not the Darwin who imputed "to nature the ugly characteristics of Victorian capitalism and colonialism" but the ecological Darwin who "moved outside the mechanical world picture" to promote an appreciation of the dynamic "continuous modification of species" (*TPOP*, 386–87). Eiseley is properly cited as an extender and intensifier of the "organic world picture." The point is less that Eiseley wrote *Darwin's Century* and *The Immense Journey* than that he foreshortened and thereby highlighted the evolutionary pace, as in his chapter in *The Immense Journey* on the "explosion" of flowers. For Mumford, "efflorescence is an archetypal example of nature's untrammeled creativity"

---

20. On the relevant behavior of schizophrenics, see Gregory Bateson, *Steps to an Ecology of Mind* (New York, 1972), and R. D. Laing, *The Divided Self* (London, 1960).

(*TPOP*, 381). And the essence of his organic world picture—or more fittingly, his organic model—is the literal and figurative efflorescence he generous-mindedly acknowledges. The power system, confusing affluence with human plenitude, tends toward automatism and away from human plenitude, failing to recognize that "automation lies at the beginning, not the end of human evolution" (*TPOP*, 399). Technocrats, Mumford perceives, know dangerously little history.

Less clearly, but I think no less surely, he celebrates Darwin as a hermeneuticist of science. Though Mumford does not use the word *hermeneutics*, its idea is implicit in his powerful generalization that Darwin set aside the usual scientific mode of isolating the part—an approach that rests on the touching faith that analyzed bits will somehow recohere—in favor of the global mode, that is, of attempting to know the whole in order to know the part. Mumford accordingly revises one of his dark fathers, H. G. Wells, who "observed, correctly enough, that mankind faced a race between education and catastrophe [but] failed to recognize that something like catastrophe has become the condition for an effective education" (*TPOP*, 411).[21] Accordingly too, almost tautologically, since *catastrophe* means "down-turning," it will take global stress to engender global reformation: what will be needed is "something like a spontaneous religious conversion . . . that will replace the mechanical world picture with an organic world picture" (*TPOP*, 413). The relatively static associations of the very metaphor of a picture betoken the prophet's difficulty in standing clear of the presiding myth he would deconstruct or replace with turning (upward) and building (*conversio*, reform).

The same sort of difficulty threatens to disable the curious epilogue, "The Advancement of Life," which begins in a midden of rancor and metaphor: he writes of "technodrama," "radioactive mind-energy," "ordeals," "channeling," "clinging" "nourishing," the "fatal weakness of all religions," "fusing . . . aspects" (*TPOP*, 415–20). But *mirabile dictu* he articulates a potent reconceptualization of steps in an ecology of culture—steps intimated throughout the two volumes but not earlier explicated.

21. For Mumford's long and various concern with education, including early school-board service, see David R. Conrad, *Education for Transformation: Implications in Lewis Mumford's Ecohumanism* (Palm Springs, Calif., 1976).

Those involve the two-way process of materialization and ethereal-ization (*TPOP,* 422–29). An *apparition* (minimally cerebral, mainly bodily/perceptual) becomes *incarnation* (in ideation conveyed by personal presence or intertextually), becomes *incorporation* (in a community or a party of believers), becomes *embodiment* (in the man-made environment). The contrary process, etherealization, is perhaps not quite so parallel as Mumford suggests. It certainly is not what Arnold Toynbee means by the term or what Mumford himself tends to mean by it in *The City in History*—a useful meaning for which he later substituted the term *de-materialization:* a bank, say, etherealizes in that sense from being a fortresslike building housing a massive vault, to being a glass-walled house for "books," to being primarily a terminal of electronic impulses and their miniaturized storage. Etherealization in Mumford's profounder sense is the returning to mind (*cf.* "returned to committee") of embodied cultural elements for rematerialization. The first stage may be physically iconoclastic: one thinks of "bare, ruin'd choirs," of the Bastille, of Stalin's statue in the Prague Spring, of the evacuation of Phnom Penh and the massacre of a great segment of Kampuchean society by Pol Pot. These examples, most of them not Mumford's, are shockingly various. But etherealization may also be iconotropic rather than iconoclastic: the DuPont home Winterthur becomes a busy commercial museum and sales enterprise; countless pretentious city houses become physicians' and attorneys' offices.

The point of which Mumford is intensely conscious is that of individual reorientation, familiar in Christian prayer asking that the sinner may "turn from his wickedness and live" and most alive to Rebecca West in the spectacle of millions of suffering refugees turning to flee[22] but observable here as the "personal experience that it is far easier to detach oneself from the system and to make a selective use of its facilities than the promoters of the Affluent Society would have their docile subjects believe" (*TPOP,* 433). In closing with the recipe of "quiet acts of mental or physical withdrawal" that "will liberate" anyone "from the domination of the pentagon of power," Mumford reads remarkably like Milton's angelic counsel, late in Book XII of *Paradise Lost,* that one may find a "paradise within," or indeed like the rejection by Milton's Jesus of the satanic offer of kingdoms of the world, even kingdoms of the mind.

22. Rebecca West, *The Meaning of Treason,* Epilogue.

Like Milton, too, is Mumford's ambiguity whether those so liber-
ated, those so reimparadised, constitute a *scherith*, that is, a remnant
that will leave the mass, or an *ecclesiola*, that is, a tiny true church that
will *not* reform the great terrene city surrounding it.[23] The very ambiguity is consistent with Mumford's final, incompletely articulated dis-
crediting of the myth of the machine. That myth compulsively seeks and
irrationally assumes predictability, a concomitant of control over time.
Mumford knows not only that "genuine novelty is unpredictable" but
also that the permutations of the given over time will be radically unpredictable and may be bizarrely divergent. His own example in witness of
that is "Jesus Christ, the most spontaneous and informal of personalities," who was yet a necessary, though emphatically not a sufficient,
cause of the cathedrals at Durham, Chartres, and Bamberg and of the
"Holy Inquisition!" (*TPOP*, 425).

Mumford is placing history against myth—the world of necessary
causes over that of mechanically deterministic sufficient causes—and is
attempting to restore an attention to the world of "phylogeny and social
inheritance (memory and history) . . . and . . . purpose" (*TPOP*, 235,
cited above, p. 97). History, as he knows, never repeats itself at the level
of atomic particularity or mythic generality but *may* repeat itself at the
level of local social patterns or corporate or civic institutional patterns.
But that history, so needful to technocrats and their fellow citizens,
must be enriched by meditation on "cultural achievements that marked
the successive generations" (*TPOP*, 13, cited above, p. 90). By that he
certainly intends, in the student of history, at least an enlivened sense of
human *presence*, of the systolic and diastolic interplay of *generations*,
and of all that not as a formless slurry but rather as *marked*.

One of Mumford's weaknesses is that he often reads with Dreiserean
heaviness rather than with the Rousseauistic verve he sought. And he
may have been oversanguine in assuming his signs to be already shared
by, one can put it, the college-educated person of goodwill whom he addresses. But he came to know in his last books that those signs must not
be reductively conceived as Saussurean as-if-atemporal ovals, as linear
emblems each with a linear bar between signifier and signified. If the
people were to *see* a sign, Mumford had to draft an emblematic figure
not in lines of closure but in a configuration of hachures—each tick of
the hachure suggesting the writing subject's temporal implication in the

23. James Hardy clarified the distinction for me.

successive generations. But even such a figure could only suggest. In culture, as in nature, the reciprocal, minutely local actions of novelty may eventuate in global effects. The *ecclesiola* might yet be a *scherith*. The prophet of Amenia might in some inconceivable yet possible train of consequences foster the reconstitution of America.

3

COLONEL TATE, IN ATTACK
AND DEFENSE ❧

*How may the Southerner take hold of his tradition? The
answer is: by violence.*

— Allen Tate, "Religion and the Old South"

*It's all very well to tell the Colonel to up and at 'em, but
fortunately the Colonel, who has a fair notion of the force
he is about to exhibit, isn't so confident that he will defeat
the enemy. He is confident that he has the right plan of
attack, but his tactics may be bad. He feels genuine ti-
midity, and if it weren't arrogant to say so, humility.*

— Allen Tate to John Peale Bishop, October 25, 1935

*Last of all, don't apologize for your "rudeness." My own
delights me.*

— Yvor Winters to Allen Tate, February 28, 1927

Mumford has written in *Sketches from Life* that he
was essentially a metropolitan until his teenage
years, and that it was later still that he came to
have intermittently, and then through the years of his last books, a full-
time home in the country. It seems fair to suppose that for the actual
metropolis in question, New York, at least such other eastern port cities
as Boston or Philadelphia might have served equally, and not impossibly
other great American cities with connections to the world at large—
Chicago, say, or San Francisco, but hardly New Orleans, as I will argue.
Mumford worked out, virtually wrestled out, an American *modus vi-
vendi* in between city and country, in between a loving appreciation of
the city and a detestation of its coercive automation, in between a loving
appreciation of a rural hamlet's rewards and an inability to tolerate its
limitations. If Platonic man is essentially—I am led by Lewis Simpson

and Eric Voegelin to the *Philebus* and *Symposium* for this point—man *in between* life and death and in between ignorance and wisdom, then American man and woman are almost sure to be also in between city and some phantasm of country, if only suburban, as well as more largely in between the Old World and the New, which is to say, between the burden of history and the temptation to quasi-atemporal world building (the frontier that has *not* closed).

Allen Tate, in apparent contrast, came from no generic metropolis, indeed from no metropolis at all. Nashville was not that; still less was Winchester, Kentucky, or Fairfax County, Virginia. And in any case Tate was not *from* Nashville, after college at Vanderbilt, in the way that Mumford was nurturantly out of New York. In the twenties, it was rather for Tate more like—in José Ortega y Gasset's apposite words about the turns of literary history—a cat from a house where it had been scalded.

Moreover, Tate was from no vaguely generic region but, on the contrary, from the South. For all the South's considerable diversity, and for all the difficulty observers have in characterizing it—no one has done better than Tate—it is by common consent the most markedly characterized region in a nation where, in the common perception, one not very profoundly idiosyncratic region shades into another. Tate, always partly a southerner, and in his last years even of Sewanee and Nashville again, came from the South fighting. He fought almost everything except his dearest friends, and even there edginess occurred from time to time. He stood *in between* in all the senses I have identified for Mumford, but in addition he stood existentially if not geographically—as if in fighting crouch—between South and North, between history and taxonomic mechanism, between mystery and "positivism," between correspondingly articulated modes of discourse (and literary and critical schools), between home and not-home. Little wonder that he has often been associated, as a critic or New Critic, with "tension," however he rationalized that term as a pun on the technical logical terms *intension* and *extension*.

His own sense of the clash of mighty opposites, of the worse prevailing, and of himself as an American caught between, came into focus very early, and never more succinctly than at the end of his brilliant and exasperated biography of Jefferson Davis:

> The South was the last stronghold of European civilization in the western hemisphere, a conservative check upon the restless expansiveness of the in-

dustrial North, and the South had to go. The South was permanently old-fashioned, backward-looking, slow, contented to live upon a modest conquest of nature, unwilling to conquer the earth's resources for the fun of the conquest; contented, in short, to take only what man needs; unwilling to juggle the needs of man in the illusory pursuit of abstract wealth. . . . The War between the States was the second and decisive struggle of the Western spirit against the European—the spirit of restless aggression against a stable spirit of ordered economy—and the Western won.

In a sense, all European history since the Reformation was concentrated in the war between the North and the South. For in the South the most conservative of the European orders had, with great power, come back to life, while in the North, opposing the Southern feudalism, had grown to be a powerful industrial state which epitomized in spirit all those middle-class, urban impulses directed against the agrarian aristocracies of Europe after the Reformation.

Tate wrote those words on Bastille Day, 1929, in Paris![1] He thought further about this in the decades following, and rather differently of the "agrarian" and of "aristocracies" (see below, pp. 145–49). But to see Union and Confederacy as synecdochic for larger cultural and historic figures—Western and European traditional or the like—remained characteristic. To understand what he gradually made for America and himself of these clashes, and their resulting and emergent figurations of reality and power, requires considering a disparity to which he, like prevailing American scholarship, was almost blind until the fifties or sixties: the difference between oral culture and literate culture. Highly literate scholars could scarcely remember their preliterate years or imagine the difference of what they were from what they had been, and citizens of orality could not tell about it in the discourse of literacy. Tate's sensitively imaginative situation in regard to the discrepancy between oral and literate—more marked in the South than in any other American region—is in more than one way ironic. And it is, I think, the place to begin with the self-proclaimed citizen of the republic of letters, because although he became extraordinarily, indeed classically, educated and a kind of prophet of the written word, he began as much an oral person as everyone else. He became precociously an omnivorous reader, but that was during a "peregrinative childhood" in—to repeat—the most residually oral cultural landscape in America. His twelve- or thir-

---

1. Allen Tate, *Jefferson Davis—His Rise and Fall: A Biographical Narrative* (New York, 1929), 301–302. The account of writing and finishing the book is in *Mem.*, 55–56.

teen-year-old facial and cranial features were palpated by an ancient, blind, sibylline black woman (but "blood cousin"), who pronounced, "He favors his grandpa"—meaning either his great-grandfather or simply, in "aristocratic courtesy," that he belonged (*Mem.*, 13). The examination here will be preparation for turning, in the following chapter, to the vexed implications of Tate's deepeningly prophetic vision for his views on the aristocracy, on history, and on authority. Ironies appear there, too, because—roughly—the more historical, the less centered is Tate's vision. Only after that, I have come to believe, will it be possible to set in clear light some of the fundamental lineaments of Tate's way with the sign and of his prophetic witness against the myth of the machine.

The bellicosity of the epigraphs at the start of the present chapter seems to have been well established in Tate's inner circle of friends. His published correspondence with Donald Davidson, for further example, shows him saying, "In this kind of literary warfare . . . you can't *prove* things like *revenge*, etc; you have to divine them and let loose. And what is life without war?" (*DD*, October 11, 1924). And the letter to "General D. Davidson," signed "A. Tate (Colonel)" (*DD*, January 18, 1933), is paralleled later in urging some secrecy and surprise in literary efforts.[2] Sometimes the military figure will be understood metaphorically, but perhaps more often metonymically, given Tate's deep sense that all language is action, a sense he shares with oral societies. In any case, the combativeness—"Bellum omnium contra omnes," he jauntily quotes Hobbes in the letter of October 11, 1924—the bellicose strategizing so characterizes much of his working style that one longtime friend standing by his grave exclaimed, "Colonel Tate!" and was instantly understood.[3]

2. *A. Tate* perhaps echoes in a reflexively mock-heroic vein the familiar *A. Lincoln.* Preparing a special issue of *American Review*, Tate wrote of "fine poetry, backed by some aloof criticism," adding that the "strategy calls for secrecy. We want this number to take 'em by surprise" (*JPB*, December 14, 1933). His language suggests an infantry attack with support by artillery.

3. Lewis Simpson, "The Critics Who Made Us: Allen Tate," *Sewanee Review*, CXIV (1986). Professor Simpson kindly let me see this in typescript.

What the essays and reviews, and even the poems, suggest about their author by their frequent edginess is almost palpable in the fifty-seven boxes of correspondence at Princeton. He wrote letters "two evenings a week, at least," in 1924 in New York (*DD*, December 17, 1924), and surely at an equivalent rate in ensuing decades. One finds a campaigner for literature, who tirelessly planned, exhorted, acted as liaison, encouraged with practical instruction, and gave reinforcement or a touch of Allen in the night. The bellicosity and sense of besiegement constitute a curiously neglected aspect of the Tatian calling to be a man of letters in behalf of the "republic of letters (which is the only kind of republic I believe in, a kind of republic that can't exist in a political republic)" (*JPB*, December 23, 1933). Lewis Simpson, in his definitive series of expositions of that calling and that republic has persuasively emphasized the quasi-priestly aspect of it.[4] Other criticism has emphasized Tate's poetic obscurity, or his personal kindness and charm, or the defects of the New Criticism in general.

Yet what we might echo Walter Ong in calling the agonistic aspect of these writings opens into another realm of and about discourse. The other realm can tell us provocative things about Tate's situation and the world he inhabited—not simply the cultural world of the South or New York or Europe but his life world and his posture within it. That conflicted and combative posture, internally and externally both allied and adversarial, is my subject here. What Tate calls his plan of attack and his tactics, which is to say, the shape of his war and its battles, his targets and his verbal ammunition, may be held aside for separate treatment.

The agonistic tone of Tate's characteristic expression has some psychic roots in family, roots of a quite common American kind. It seems necessary to acknowledge as much, and more might be said.[5] But cul-

---

4. See *ibid.*, and Lewis Simpson, "The Southern Republic of Letters and *I'll Take My Stand*," in *A Band of Prophets*, ed. William C. Havard and Walter Sullivan (Baton Rouge, 1982); Simpson, *The Dispossessed Garden: Pastoral and History in Southern Literature* (Athens, Ga., 1975); and Simpson, Foreword to *The Literary Correspondence of Donald Davidson and Allen Tate*. On Tate's literariness in contrast to Davidson's bardic championship (albeit in his essays!) of the spoken word, see Simpson, Introduction to *Still Rebels, Still Yankees, and Other Essays*, by Donald Davidson (Baton Rouge, 1972), v–xvi.

5. In 1972, in "A Lost Traveller's Dream" (*Mem.*, 3–23), Tate described such notes as from a lived-in forest of the self which is a *selva obscura*, between fiction and history. But, "until I was thirty I never lived in one place longer than three years" (he resembled Eiseley in this), and he had a chilly father who shamed him by public scolding, by social disgrace, and by economic failure. He had an overcontrolling mother who told him his mind was

turally, as Ong has shown, Tate's tone of expression associates with a congeries of features that distinguish oral from literate culture. The former contrasts with the latter, Ong argues, as (*a*) additive and aggregative, rather than subordinative and analytic, (*b*) redundant and copious, rather than dense or concise or allusive, (*c*) continuous, rather than intermittent, (*d*) conservative and traditionalist, rather than innovative and eclectic, (*e*) close to the human life world and situational, rather than abstract and static, (*f*) agonistic and vituperative, or equivalently fulsome in praise, rather than reportorial and judicious, (*g*) empathic and participatory, rather than objectively distanced, (*h*) homeostatic in adjusting history to current felt realities, rather than receptive in seeking to perceive a historical occasion as if through contemporary eyes (however awkwardly the lens might end up being handled on particular literate occasions).[6]

It must be added that Ong's work not only instructs as to the differences in kind between oral cultures and literate cultures and reminds that in literate cultures orality will be variously residual and "secondary," which is to say, framed and altered by literacy, but it also alerts us to the degrees of difference between or of intermixture of, literate and oral—and to the persisting degree, since every infant is without a word and may be reared by circles of people whose thinking and expressive behavior are *more or less* habituated to the protocols of literacy. These considerations could well frame separate treatments of Tatian issues in terms of other focuses of critical theory.

Still, as conspicuous to Tate's readers as his combativeness is his hypertrophic literacy. All of the features Ong ascribes (etymologically, "writes down for") to literacy and many he ascribes to orality appear over time in Tate's diction, his syntax, and his organizational habits beyond the sentence. In writing, Tate analytically emplaces countless references to other texts and other analyses. This is true even in the relatively conversational letters. But that may be less of a paradox than it at first seems. For one thing, writing so as to sound (sound?) conversational is a

---

not very strong. He *dreamed* unforgettably of Uncle Leo, "ruined by a powerful matriarchal mother," blocking the boy from getting home to the top floor of an apartment house: did the Dark Father/Surrogate Mother/Surrogate Self stand in the way of maturity?

6. See Walter J. Ong, *Orality and Literacy: The Technologizing of the Word* (New York, 1982), esp. 37–57. Ong's ideas are used here with some adaptation: his notes indicate more extended treatments in his own and others' earlier work.

literary skill of marked sophistication, as any reader of college under-graduate papers knows.

The persistence of traits of orality—and not only the agonistic or panegyric—in Tate's elaborate literacy can seem more profoundly para-doxical: his southernness, his espousal of values that he *called* essen-tially southern, may be conceived as strong affection, despite its ambiva-lence, not only for native hearth (in Rebecca West's sense; no traitor he) but for the conservatism and immediacy normal in oral society. Those themes run through Tate's work, especially the essays with a strong so-cial focus—he would have bridled at calling it *functional*—from "Reli-gion and the Old South," in 1930, to "Faulkner's *Sanctuary* and the Southern Myth," in 1968.[7]

The first was Tate's contribution to *I'll Take My Stand*.[8] Tate, Andrew Lytle, and Robert Penn Warren objected to that title for the collection (*DD*, Appendix D), apparently feeling it suggested unreflective rigidity perhaps worse than bellicosity. In any case, Tate and some others who had made *The Fugitive* a significant little magazine in the early mid-twenties had exhibited, in Simpson's words, something like the "pattern of withdrawal and return common in the psychology of visions."[9]

Louise Cowan quotes his letter to Marjorie Swett of 1923: "We fear very much to have the slightest stress laid upon Southern traditions in literature; we who are Southern know the fatality of such an attitude—the old atavism and sentimentality are always imminent."[10] Such damn-

---

7. As early as February, 1927, Yvor Winters praised Tate's analytic brevity in pieces for the *New Republic* and the *Nation;* as late as 1943 Tate wrote in "The Angelic Imagi-nation" that "insights of the critical intelligence . . . may never wholly rise above, the subtle and elusive implications of the common language to which the writer is born" (*Col.*, 447).

8. *I'll Take My Stand*, by Twelve Southerners (1931; rpr. with introduction by Louis D. Rubin, Jr., Baton Rouge, 1977).

9. Lewis P. Simpson, *The Man of Letters in New England and the South: Essays on the History of the Literary Vocation in America* (Baton Rouge, 1973), 244.

10. Louise Cowan, *The Fugitive Group: A Literary History* (Baton Rouge, 1959), 116; Simpson, *The Man of Letters*, 245.

ing words as *atavism* and *sentimentality* were not uncharacteristic of his ambivalence. The next year he would not return south from New York, he said, unless a job Warren was trying to secure for him became definite: "I simply can't pursue an *ignis fatuus*, especially in the South, a region none too friendly at best" (*DD*, June 5, 1924). By a small irony—one surely not without counterparts in his voluminous correspondence—this crossed in the mail with Davidson's complaint that a review by Tate for the Nashville *Tennessean* used words that no more than half a dozen among the paper's readership would understand. By Christmas, Tate was saying with truth enough but more than expectable detachment: "I can never forget you all. But really I shall never return to Nashville; so you must come up here when you can" (*DD*, December 17, 1924). Cowan quotes William Yandell Elliott, who wrote from Berkeley in 1923 of "Tate, that child of wrathful detachment." [11]

Somewhat earlier, already in New York, he had pursued with his uniquely deferential assertiveness a corporate definition of a proper Fugitive relationship to home, and the relationship is a written one:

> It isn't the old South as material we object to, it seems to me (all Greek literature is a throwback to a fragrant and heroic past), but the fatal attitude of the South toward this material. . . . There's nothing wrong with local color, so far as I can see, except when it drops to mere colored locality—everything must be placed in space and time somewhere, and the South is as good a correlative of emotion as any place else; and so I think that the trouble is the damnably barbaric Southern mind, which would be provincial in London, Greenland, or Timbuctoo. (*DD*, June 29, 1923)

Hoi barbaroi the Greeks had called the tribesmen from the north and west of them, by a metonymy from how their speech sounded to the Greek ear. Of all Greek literature, that which is notably a throwback to the "fragrant and heroic past" is not the textual poetry of the Greek Anthology or the works in what Ong points out to have been the first text-bound genre, the drama, but rather the epic, the more oral and tribal form. The point is that the ethnocentricity of orality and of tribalism go together as the convex and the concave of a curve, and Tate seems to have thought something like that ("fatal attitude") even as one exiled by his literacy from the epic garden. *Placing* in space and time is textual or quasi-textual. Most of Tate's visits, likewise, were epistolary.

Moreover, raising the whole constellation of issues to consciousness,

11. *Ibid.*, 131.

so as to achieve an awareness of options, precisely an awareness *not* provincial, accords as paradoxically with the nature of the object as a map of chaos would accord with the "spirit of the place" (to use Paul Pickrel's deft trope). In a further irony that must have been uncomfortable, Tate partly recognized that, too. After publication of *I'll Take My Stand,* he wrote to John Peale Bishop, on the matter of relationship to the land, that "like most of us you are both inside and outside the old tradition, . . . in a word you are a modern and divided mind" (*JPB,* June, 1931). Whatever the justice of Tate's objection to the title, the text did, in contrast to waves and counterwaves of oratory, *take its stand.* As Ong prompts us to keep in mind, any text is intransigent. Any text answers questions no more than does a Grecian urn, even though the author may do so in subsequent voice or text. R. P. Blackmur may have had some such in mind as he wrote of Tate's standing in a "kind of violence of uncertainty and conviction, deeply entangled in his beliefs and absurdly outside them." [12]

"Religion and the Old South" is in considerable part about a horse. Tate, as if donning the "overhauls" of village storyteller, copiously propounds a parable, albeit disjointed, of the "whole horse" of viable religion living in the meadow of experience, comparable to Ortega's "lived-in forest" of *Meditations on Quixote*—a situation neither idealist nor realist. Tate mainly distinguishes the whole horse from the rationalistic, goal-oriented, and properly mechanistic half-horse of mere horsepower, secondarily from the feckless absurdity of irrational "symbolist poets and . . . Bergson."

But Tate leads simple narration some intricate measures—with analytic and ironic differentiations of (in Clifford Geertz's term) "thickly historical" Christ from abstractly historical Osiris, and of horse as Christ from centaur—before he lets the argument subside into unevenness and assertiveness in the last two of the essay's six sections. There may well be a self-consciously ironic relationship of intertextuality between Tate's horse and the Emersonian one-horse shay of Calvinism. (Tate wrote of a friend's fictive characters that "it is the irony which testifies that the author has faced the present [in relation to the past]"; *JPB,* March 12, 1931.) But more assuredly Tate here concerns himself with the horsepower of northern reductive instrumental thinking. And I

---

12. R. P. Blackmur, "San Giovanni in Venere: Allen Tate as Man of Letters," in *Allen Tate and His Work,* ed. Radcliffe Squires (Minneapolis, 1972), 102.

am concerned at this point not with his conception of South, North, Bergson, or Jefferson but rather with his attitude toward conceptions he could not live in as reality yet could half approve as myth, and the simultaneous closeness and distance of his peripatetic self with regard to southern neighbors.

Tate begins the essay as if mediating between anthropologists in residence and some more or less transient third parties, speaking of his own religion as "fable" and "myth" but religion all the same, "immediate, direct, overwhelming" (*Col.*, 306). After defining the religion of the whole horse by contrast with those of the half-horse, in an argumentative strategy empathically—or else ironically—close to an agrarian life world, he turns to analytic distinctions as highly abstract as anything he attacks for abstraction. The "religion of the completely workable" is truly a religion, he argues, despite its deficiency in concretely situational and participatory markers like altars and rituals, because it *worships* something it defines as an absolute: logical necessity and human rationality. He appeals to the relatively oral life world of experience, on the one hand, to assert the inadequacy of that, and to the literate abstractive device of the *reductio ad absurdum*, on the other hand, to falsify the idolized rationality: "It can predict only success" (*Col.*, 307–308). As rhetoric for quarreling with others, the approach risks self-defeat; as the poetry of quarrel with self, it is poignantly suggestive (Yeats's distinction, endorsed by Tate in *Mem.*, 146).[13]

Recognizing limitations in the "image of the horse" and seeking lifeworldly immediacy, Tate proposes a supplementary double "image of history," allegorized, and taxonomized, as a potentially Christian short view treating history as a "concrete series" with the same "accident and uncertainty" as the present, and a long view treating it as a "logical series," with general identities and rationalistic "natural law," the "cosmopolitan destroyer of tradition." It is not to my purpose to quibble here with his fairness in characterizing historians or with the local details of his logic. Both historical attitudes he identifies have been discernible at large; Tate's view of his South never fails to recognize the spiritual deadliness to whites of slavery or the physical murderousness to blacks of slavery and postemancipation racism. But his epistolary friend Yvor Winters took him immediately to task in a letter, and others did

13. See the partly similar argument of Michel Serres in his *Hermes: Literature, Science, Philosophy*, ed. Josué Harari and David F. Bell (Baltimore, 1982), 15–28.

publicly, for the vaguely desperate announcement at essay's end that only violence could now enable the southerner to take hold of his tradition.[14]

Still less would I take issue with Simpson's provocative and consequential point that "Ransom, Davidson, and Tate . . . yearned, it might be said, to complete the spiritualization of the secular. By transforming the South into a symbol of a recovered society of myth and tradition, they would assert the community and spiritual authority of men of letters and make whole the fragmented realm of mind and letters."[15] But Tate himself insists on a "nice and somewhat slippery paradox here," and it is one that might trouble anyone who from the universal illiteracy of infancy achieves unusual literacy. Even if one's literacy, in all the duality of oral-literate capabilities anatomized by Ong's schema, were not so superbly developed as Tate's, one's own society would be modified, as Ong indicates, by one's capabilities and their doubleness. Even if other things could be equal, a child coming to literacy in a home with literate parents would have differing nurture according to whether *their* parents were literate or not; the degrees of spatializing, analytic thought would differ. The "tension in poetry"—referred to in his article with that phrase as title, and elsewhere—between Platonist, abstractive (ex)tension and connotative, ambiguous (in)tension, seen as a *scale* along which "the poet deploys his resources," looks to be in part an orality-literacy tug-of-war (see *Reas.*, 76).

Tate pursues the matter as if in oral debate, by postulating the question, Why must we choose whole horse or half, short view or long? "Merely living in a certain stream of civilized influence does not compel us to be loyal to it. Indeed, the act of loyalty, even the fact of loyalty, must be spontaneous to count at all. Tradition must, in other words, be automatically operative before it can be called tradition" (*Col.*, 311). He closes the section by acknowledging a little ruefully the paradox that both tradition and the long view prompt the "present defense of the religious attitude." He finds "this conception . . . wholly irrational" and calls irrational his own efforts "to discover the place that religion holds, with abstract instruments, which of course tend to put religion into some logical system or series, where it vanishes" (*Col.*, 311–12). Less

14. Yvor Winters to Allen Tate, December 29, 1930, in Tate Collection, Firestone Library, Princeton University. Quoted by permission of Janet Winters.

15. Simpson, "The Southern Republic of Letters," 87–88.

hard on him than he, we might define the efforts as more ironic than irrational: the ultraliterate both defining and not defining tradition and religion in terms of the modes of oral culture—almost in terms of the Edenic garden from which his knowledge excludes him.

Yet religion did not vanish for Tate himself: he eventually became a Roman Catholic. Is he making an asseveration that secular means cannot spiritualize the secular? Ong argues that "text-formed thought" about literacy, orality, and identity can—not will but can—deepen a religious sense of the human situation. Was Tate, then, confused or consciously ironic at age thirty-one? "Religion . . . vanishes" means that it vanishes for anyone ill advised enough to worship reductive rationality. But *how* situate religion with text-formed habits of thought yet not *place* it "into some" reductive quasi thing? Can "religion . . . vanishes" mean, with a milder and more compassionate irony, that it suffers some troubling attenuation? Something of that, though Tate's scorn is real for those who would, as we have learned to say, lapse out of history and back into myth, whether the myth of science or politics or history. He seems generally in step with a half-century of Continential argument in behalf of history over myth, most of it still to come when he wrote the essays first treated here, and with Continental phenomenologically existential appraisals of lived experience, including religious experience (not Edmund Husserl, Jean Paul Sartre, and Albert Camus, but rather Karl Jaspers, Gabriel Marcel, Han-Georg Gadamer, and Eric Voegelin).

Some of his diagnosis presages a good deal of later discourse. He appraises the characteristic early-nineteenth-century New England attitude as "self-conscious and colonial," attributing to it an outlook that is possible only in a society considerably conditioned by text-formed thought, in contrast to the South, which seemed to itself, in the usual attitude of oral culture, "self-sufficient and self-evident." He goes on: "The Southern mind was simple, not top-heavy with learning it had no need of, unintellectual, and composed; it was personal and dramatic, rather than abstract and metaphysical; it was sensuous because it lived close to a natural scene of great variety and interest. . . . They liked very simple stories with a moral in which again they could see an image of themselves" (*Col.*, 319–20). An *alas* just before those lines betokens the Virginian, Kentuckian, and Tennessean Tate's ironic awareness that the southern mind had in fact needed more learning than it recognized. Myth, as he eventually came to define it, means "a dramatic projection of heroic action, or the tragic failure of heroic action, upon the reality of the common life of society, so that the myth *is* reality" ("Faulkner's *Sanctuary* and the Southern Myth," in *Mem.*, 151). His bit about "simple stories" is the decisive diagnosis, even if he does not quite have the decisive terminology. Ong argues that past ages were not ontologically more heroic than this one but that, rather, aural apprehension and mnemonic economy in an oral society *demand* sharply differentiated type characters, the oral-aural equivalents of the white hats and black hats in the early Western movie. Northrop Frye has long since anatomized the Western as a version of pastoral, and it is notable that pastoral seems historically to have thrived in connection with secondary, that is, partially text-transformed, orality. Narrative, Ong observes, "builds less and less on 'heavy' figures until, some three centuries after print, it can move comfortably in the ordinary human lifeworld typical of the novel." [16] To say it can is not to say it regularly does. Tate's words, rather than Ong's, well describe the world of most television serials, soapy and otherwise, a world evidently appealing to minds not much or not long affected by text-formed thought. Likewise, whatever the "southern gothic" owes to action in southern life worlds, it owes something also to the oral traditions about heavy characters. Elsewhere Tate praises Bishop's literately ironic treatment of quasi-historical type characters (*JPB*, May 11, 1931).

Simpson, in his foreword to the Tate-Davidson letters, sums up one line of argument and Tate's own sense of vocation by saying aptly that he sought to "act heroically by becoming a man of letters committed to changing the world view of modernity" (*DD*, ix). Just so: *heroically*, rather than, say, *expertly*, marks the conflictual persistence of orality. [17]

Somewhat similarly, one should consider Tate's reflections on the plight and deformations of the sign, from strictures in 1930 about reason and

---

16. Ong, *Orality and Literacy*, 70.

17. Radcliffe Squires sees Tate's thematic concern "shifting from 'failure' toward 'vision'"—from tribal agonistic to textually cool, one might say (*Allen Tate and His Work*, 7). But Tate was *deliberately* truculent early (Cowan, *The Fugitive Group*, 115*n*36) and late (he would keep a feud alive; Allen Tate to Charlotte Kohler, at the *Virginia Quarterly Review*, February 23, 1945, in Tate File, Alderman Library, University of Virginia).

nature being wrongly equated in what was called naturalism (*Col.*, 314) to "Ezra Pound and the Bollingen Prize" (1949) and later. One should consider Tate's concern with uniting man's "moral nature and his economics . . . that greatest of all human tasks" (*Col.*, 302). One should consider Tate's campaign for "a spiritual community . . and for the mind's capacity for the perception of transcendence," and one profits from Simpson's having done so in his foreword to the Davidson letters and since. But more remains to be said about Tate's continuing struggle, increasingly conscious and informed, with the modes and the issues of orality and literacy. If Ong has been the foremost analyst of those issues, Tate may thus far, I think, be called one of the most noteworthy American protagonists.

He wrote "What Is a Traditional Society?" early in 1936, thought it the "best essay" he had "ever done . . . in the field of general ideas" (*JPB*, March 31, 1936), gave it as the Phi Beta Kappa address in June at the University of Virginia, and published it in the *American Review* in September. Urbanely—or homeostatically?—adjusted to the audience at "Mr. Jefferson's university," it works the crowd superbly, from the most puerile undergraduates to their most senior mentors. It is a remarkable mixed work. In it, one finds "finance-capitalism" posited as a five-day-a-week way of life at odds with the potentially more humane and spiritual way of life often pursued only two days a week. Indeed, Tate avers that "finance-capitalism, a system that has removed men from the responsible control of the means of livelihood, is *necessarily hostile* to the development of a moral nature" (*Col.*, 303; my emphasis). One can only infer here what he elsewhere makes explicit, his religious objection to any practice that treats persons as things, which he finds doubly objectionable in the case of the black man, who is both made a thing and so situated as to alienate the slaveholder, who degrades him, from the land that might in Tate's view be the medium of the slaveholder's salvation (*Col.*, 273). One gets in this address essay an oratorical expansiveness, and devices like anaphora difficult to represent briefly, though "finance-capitalism" hints at the copiousness of orality (*cf.* the still-current southernism *toad-frog*).[18] But one does not get an

18. See Stephen M. Ross, "Oratory and the Dialogical in *Absalom, Absalom!*" in *Intertextuality in Faulkner*, ed. Michel Gresset and Noel Polk (Jackson, Miss., 1985), esp. 75, 78. Ross cites Faulkner as characterizing the style of *AA* in "studbook style: 'by Southern Rhetoric out of Solitude.'"

explication of the objections that can be drawn as implications from religious belief, and those might well elude a listener. A sufficiently analytic listener in 1936 might, like a reader today, flinch at the pastoral neatness of the claim that Virginia gentlemen of the early Republic "knew what they wanted because they knew what they, themselves, were" (*Col.*, 297); as we shall see, Tate later disavows the ascription of self-knowledge. Or the reader might fix on his stipulative definition: "An untraditional society does not permit its members to pass to the next generation what it received from its immediate past" (*Col.*, 302). *Pass, what, it*—Tate's reified abstractions may seem incautiously unitary. Does not the implied opposite, traditional society, resemble at this point in his essay homogeneous, oral, tribal society? We know that orality persists powerfully in a moderately literate culture—as for a Tatian example, the many more farmers' sons among the people of the Book who became farmers than preachers. We know, with Tate, that the powerfully more-than-literate nexus of family may override traditional *or* high literate culture—can pass a crazy salad whereby the horn of plenty is undone. We know, as he perhaps more guessed than knew, that all the overlapping ways of cultural transmission—family attitudes and exempla, family emotional and economic conditions, semipublic experience, public exempla, formal education—will be rearticulated in the wake of the previous generation's literacy.

His thought and behavior already suggest a campaign in behalf of the "republic of letters." He had taken at Southwestern University, in Memphis, the first of several temporary teaching posts. With Herbert Agar, he was editing *Who Owns America?* a collection of essays to be a textbook campaign manual for enlistees to the cause. And he exhorts his listeners in Charlottesville and the readers of Agar's *American Review*, "Traditional property in land was the primary medium through which man expressed his moral nature; and our task is to restore it *or to get its equivalent today*" (*Col.*, 303; my emphasis).

In the thirties, Tate tended to write about land less as a metaphor or medium and more as literal soil, as arable property. But he *wrote*, whereas his wife, Caroline Gordon, gardened and enjoyed the outdoors, as Ann Waldron has shown.[19] He wrote in 1931 that the "excessive politeness to women" in the antebellum and later South "was due to the

19. Ann Waldron, *Close Connections: Caroline Gordon and the Southern Renaissance* (New York, 1988), esp. Chapter 7.

realization that they were the material medium . . . through which property is perpetuated, and while we had no respect for them as individuals, we had an exalted respect through them for the land" (*JPB*, 36). This, again, is to define in the analytic irony—and syntax!—of literacy the deindividualizing heavy characterizing of oral culture. Again, in "The Profession of Letters in the South" (1935), he wrote that the antebellum "white man got nothing from the Negro, no profound image of himself in terms of the soil." That too marks the literate man's externality to his markedly oral home culture, toward which he has such sympathy and ambivalence. How *could* an oral-aural or at least residually oral white South get an *image* of *itself* from a commoditized black man situated between itself and the soil? Such interdiscursive imaginative and literate prestidigitation would be unlikely. That, of course, does not deny what Tate deplored, the bad effects on southern cultural life of chattel slavery, and oralistic phantasms of the black other.

What shall metaphoric land be? The formulation of any answer would seem for Tate to depend on literary and ideological efforts of men and women citizens from the republic of letters. The land when conceived as medium, tenor, or vehicle for expressing man's moral nature, insofar as land is cultivated and acculturated, abuts on such diverse phenomena as ecology consciousness, a Miltonic-Puritan sacramental sense of work, and a psychoanalytic sense of adult self-actualization, generativity, and creativity (the not precisely synonymous terms of Abraham Maslow, Erik Erikson, and Mumford, respectively). If the "equivalent" for land, metaphoric land, is language itself, the republic of letters requires higher general literacy than ever yet achieved. Without that, the republic of letters will certainly be a Gnostic cult of oligarchy (the besetting danger in any case). In "The Profession of Letters in the South," Tate partially glosses himself: his declamation against the "romanticism" of older southern literature, with its "formless revery and correct sentiment [and] inflated oratory" and its "roots . . . not deep enough" to overcome the "deficiencies in spiritual soil" (*Col.*, 274), itself suggests an attenuated sort of secondary orality.

In a kind of simultaneous example and crux of his argument and dramatization of his proposal, he quotes the "Game of Chess" section from *The Waste Land,* and glosses it. It is important to remember that he always honored Eliot as one of his two most influential teachers, John Crowe Ransom being the other. He says, in part, "In ages [like ours] which suffer the decay of manners, religion, morals, codes, our inde-structible vitality demands expression in violence and chaos; it means that men who have lost both the higher myth of religion and the lower myth of historical dramatization have lost the forms of human action; it means that they are no longer capable of defining a human objective, of forming a dramatic conception of human nature" (*Col.*, 301). *Forms,* here as elsewhere with Tate, means more than "visible shapes." It means something like "modes and principles of arrangement, similarity, and experienced continuity"—a religious or near-religious concept for him. Thus, he brilliantly uses the text-formed skills of analysis and subor-dination to attack the discontinuous and abstractive tendencies of thought that has run to atomistic extremes. He attacks in defense of "tradition," which is something continuous, participatory, aggregative, homeostatic, and at its best neither exclusive nor repressive. Is it an ab-surdity or forlorn hope for literacy to be defending orality? No. Each human being must achieve orality before achieving literacy, and the ge-nius of the former, like ideas in culture, need not die even though it is resituated. That we know our vitality is far from indestructible strength-ens Tate's point. He may be seen as a prophet and a pioneer, in both agrarian and military senses, in behalf of an enlightened secondary or-ality, one attempting to discriminate in favor of the best that has been thought and believed (to revise Matthew Arnold's phrase). Spiritualiz-ing the secular world will, formally, be textualizing it for communities necessarily highly literate. The danger, as Tate recognizes—and as the next chapter will develop it—is that the secular world may be sacralized or Gnosticized, in a reaction like that of Milton's Mammon or of Belial.

The tricky footwork and occasional awkwardness of Tate's attempt appear in a somewhat less focused essay of the year before, "The Profes-sion of Letters in the South," originally in the *Virginia Quarterly Re-view* in 1935. His reflections on the South, on letters, and on the profes-sion of letters range long in time, far in space. Still earlier, he had animadverted privately on Archibald MacLeish's *Conquistador:* "Fac-ing our own past is a way of facing the present. Has MacLeish any Aztec ancestry? It is probably too bad that a man like MacLeish ever received an education, for education is nothing without tradition, and the tradi-tionless educated man will, in spite of himself, work out fictitious and romantic scenes to dramatize his character in, usually in the past. It is a blessing that Ernest never went to a university and took it seriously" (*JPB*, March 12, 1931). Whatever one may think about Hemingway or

MacLeish, one must acknowledge that writers who have been endowed with education and tradition have frequently projected elements or fantasies of self into historic or quasi-historic scenes. Tate was later and by his own admission to do some of that in his only novel, *The Fathers* (1936).

In "The Profession of Letters in the South," he offers an extraordinary delineation of the difference of the twentieth-century South from the twentieth-century North. He begins by acknowledging an "immensely complicated region," as privately he would admit that "the 'North' is a group of sections whose interests should be severally protected from the haphazard impulse of industrial capitalism." [20] But he insists that despite sectional diversity it is a "single culture":

> The South to this day finds its most perfect contrast in the North. In religious and social feeling . . . the greater resemblance [is] to France. . . . Englishmen have told us that we still have the eighteenth-century amiability and consideration of manners, supplanted in their country by middleclass reticence and suspicion. . . . Where, outside the South, is there a society that believes even covertly in the Code of Honor? . . . Where else in the modern world is the patriarchal family still innocent of the rise and power of other forms of society? Possibly in France; probably in the peasant countries of the Balkans. . . . Where else [is there] so much of the ancient land-society . . . along with the infatuated avowal of beliefs hostile to it [and a] supine enthusiasm for being amiable to forces undermining the life that supports the amiability? . . . Its . . . religion [is] a convinced supernaturalism . . . nearer to Aquinas than to Calvin, Wesley, or Knox. (*Col.*, 268–69)

I have quoted at length, even while eliding some oratorical pavanes and divagations, because the passage instances so many of the cultural ambiguities in Tate's position as an ultraliterate southerner and shows something of his ambivalence breaking through the habit and conviction of mannerliness: "infatuated . . . supine"; compare Othello's "Goates and Monkeys."

The suggestion of cultural kinship with "peasant countries of the Balkans" is a shrewd one, however defamiliarizing. In "The New Provincialism" (1945), he formulates in his most considered fashion a

20. Tate, writing on the back of a letter from George Soule, then one of the editors of the *New Republic*, November 18, 1931, in Tate Collection, Firestone Library, Princeton University. It is not clear whether he wrote that to Soule. Published with permission of Princeton University Library and Helen H. Tate.

provocative distinction between what he calls regionalism and provincialism:

> Regionalism is that consciousness or that habit of men in a given locality which influences them to certain patterns of thought and conduct handed to them by their ancestors. Regionalism is thus limited in space but not in time. The provincial attitude is limited in time but not in space. When the regional man, in his ignorance . . . of the world, extends his own immediate necessities into the world, and assumes that the present moment is unique, he becomes the provincial man, [and] without benefit of the fund of traditional wisdom [he] approaches the simplest problems of life as if nobody had ever heard of them before. A society without arts, said Plato, lives by chance. The provincial man, locked in the present, lives by chance. (*Col.*, 286–87)

Ong has contended in *Orality and Literacy* that Plato was really arguing for the new textual literacy and against the old reductive orality embodied in poets and rhetors. Tate hedges: consciousness *or* habit. He would seduce by metaphoricity: "handed to them by their ancestors." Precisely not by hand do traditional cultures in "peasant countries" of the South or the Balkans or Africa convey the traditional messages, except insofar as a minatory hand "touched . . . trembling ears"—or the body of a recalcitrant child, or the like—to reinforce a vocal message.[21] And the vocal message tends to be more diffuse, harder to taxonomize or systematize than one that is more literate and linear. Tate praises Yvor Winters' linearity (1928; *Rev.*, 67) but also Hart Crane's truth of legend over history (1930; *Rev.*, 101).

"Tacky is tacky," remarked a well-bred southern lady of my acquaintance, "and if he doesn't know what it means, there's no use trying to explain." She is university educated but was adverting to a socioaesthetic gestalt "handed" to her by the time of her midteens and probably little addressed in any of her schooling. *Handed* is a metaphor for vocal interpersonal presence, which is why regionalism can be extended in time but not in space (unless electronic means to "reach out and touch someone" become more internalized?). In a 1933 review complaining of too little action in E. A. Robinson's long poems, Tate concedes that "with the disappearance of general patterns of conduct, the power to depict action that is both *single* and *complete* also disappears" (*Rev.*, 167). But the heavy characters and single patterns of oral culture, the

21. Again, in "The Profession of Letters in the South": "All the great cultures have been rooted in . . . free peasantries" (*Col.*, 273).

good, the bad, the tacky, can be repressively oversimple. (The tacky seems to be a formulation of secondary orality.) And their completeness and singularity, their very marked boundedness, will be eroded by the profuse variety affordable to print culture, which provides the opportunity for good or ill in print culture but sure demoralization for oral culture. Indeed, René Girard has argued in *Violence and the Sacred* that just the kind of loss of boundaries or "patterns" that results constitutes cultural crisis.

"Traditional wisdom" can bear down with an oppressively or stultifyingly heavy hand, eventuating in tribal stasis, degrading linear history into cyclical myth. Tate himself lived for years in partial flight from the South, not to the quasi-tribal society of the Yankee village or of, say, Queens, but to the New York City literary scene, which he saw as factionalized but not tribalized.

Yet the metaphor of handing on culture admits to the problem that Plato could not resolve. Handing suggests physical presence or something like voice, but alas, the text or artifact that can be literally handed over can more easily than the live though evanescent voice go forever ignored. A society is still "without arts" and the more a slave to chance insofar as neither an aggregation of culture's objects nor the community of persons who might appreciate them sojourns in the house of the muses, whether museum, library, or pedagogical adjunct to those. With neither elocutionary nor textual arts internalized by the populace as mediums keeping history alive, society does lapse into *tuchā*, chance, the atomism of tychastic time.

Tate puts it in "The New Provincialism," with acknowledgment to Christopher Dawson and to house and temple, that "man belonged to his village, valley, mountain, or seacoast; but wherever he was he was a Christian whose Hebraic discipline had tempered his tribal savagery and whose classical humanism had moderated the literal imperative of his Christianity to suicidal otherworldliness" (*Col.*, 289). That such otherworldliness was not always and everywhere made a central Christian imperative, in the way of loving one's neighbor, for instance, does not detract from the present point that the imperative tended to be literary, as in the Pauline epistles, or secondarily oral, as in homiletic argument from them. Worth emphasizing too is one shared aspect of classical humanism and Judeo-Christian traditions which could nestle comfortably enough with tribalism—and even with its predilection for xenophobic savagery—but which helped articulate Tate's opposition to modern instrumentalisms. That is the preoccupation of Hebraic discipline and, perhaps to a lesser degree, classical humanism with time, with long views and processes. In the Judeo-Christian instance, the preoccupation with extended history is partly a metaphoric genuflection toward eternity. That fact further marks (rather than attenuates) Tate's estrangement from the taxonomic atemporality of what he tended to stigmatize as positivism. Atemporality is quasi-unmediated, always quasi-local, often visualized. Eternity is infinite timelessness always mediated by an always inadequate imaginative expression.

But the prophet can remind a perhaps reluctant audience that man is the being who has conceived eternity, and who has conceived human kinships that—unlike mechanical control—extend "unchanging" over very long sequences of time. Tate calls the prophet who does that the "man of letters in the modern world," in the words from the title of an essay of 1952 that will be considered at some length in Chapter 5. Here suffice it to record only that the man or woman of letters—as Tate's championship of more than one woman writer indicates—is to be the person in between, the person mediating and distinguishing between present flux and an "unchanging source of knowledge" and between "communication, for the control of men, and the knowledge . . . for human participation" (*For.*, 4), to be, in short, the prophet.

Tate's sense of man as a being with a very long-term corporate and individual form of identity, a form more exactly describable as a stable range of possibilities, would by itself set his view in opposition to mechanicoplastic proposals for abrupt remodelings of humankind. His belief in an unchanging source of knowledge, a source by definition incompletely knowable, and his conception of man as himself incompletely knowable because a believing being jointly honor knowledge but dismiss dreams of totalization. There is a kind of high and difficult middle ground between ignorance and positivism, as he would limn the matter: "Man is a creature that in the long run has got to believe in order to know, and to know in order to do. For doing without knowing is machine behavior . . . with which man's specific destiny has no connection" (*For.*, 7). The prophet as a man of letters and hermeneuticist is obliged like Saint Luke's preacher to deliver captives of the myth of the machine, assist in the recovery of sight to those blind to theres beyond the here or to thens beyond the now or to possibilities beyond the mechanically implicit. In sum, as Tate might have been increasingly willing to say in his later career, the man of letters must argue the "acceptable year of the Lord." Such a prophet had better go armed.

"A Southern Mode of the Imagination" (1960) stands as some of Tate's ripest public and explicit address to this complex of issues. It is retrospective and revisionary, quoting "The Profession of Letters in the South" not as Matthew Arnold quoted himself, for choral reinforcement, but to gain a dialectical purchase for new and further thought. Remarkably, it was almost Tate's last substantial essay, though his mind was clear until his death in 1977.[22] It is here that he puts the finest point he can on the old distinction between rhetoric and dialectic, in order to differentiate the southern mind. The rhetorical mode of discourse, "the traditional Southern mode[,] . . . presupposes somebody at the other end silently listening. . . . Its historical rival is the dialectical mode, or the give and take between two minds, even if one mind, like the mind of Socrates, prevail at the end. . . . Southern conversation is not going anywhere . . . is only an expression of manners, the purpose of which is to make everybody happy" (*Col.*, 560). Oral societies are conservative. Strongly hierarchical social structure reinforces that conservatism; those in power like it, and the structure is so stable that even for those not content, the prospects of change can scarcely be conceived at all— certainly not as the quasi-reified options and sequelae of text-formed thought. Change is alien and tends, at least initially, to be feared. Tate has sometimes been unfairly linked with the old-line racist opposition to all change, because he opposed change in the direction of the machine myth. He never believed in that listening which is mere acquiescent waiting one's turn, nondialectical listening and speaking toward a fatuous tribal state of "everybody happy."

But, it may be objected, what about the bellicosity of oral society? What about *flyting?* What, indeed? Tate's remarks occur in a context of argument, opposing a long gray line from Robert E. Lee to Henry W. Grady, opposing in more or less temporary alliance Mark Twain, W. J.

22. Tate's short essay to introduce the New American Library edition of *Sanctuary* (1968) and a memoir (1972) are printed in *Mem.*

Cash, Ernest Hemingway, Henry Adams, T. S. Eliot, Lionel Trilling, and Leo Marx. Moreover, the verbal combats of orality are rudimentary in comparison with the warfare of linear argument and tend to forms like "You're a bigger one." The invective of orality may in Tate's and Ong's terms be seen as a subset of normal rhetoric in which copious or exaggerated expressions are exchanged by listeners waiting their chance for self-expressions that need not connect. Tate elaborates: "The Southerner always talks to somebody else . . . but the conversation . . . isn't going anywhere; it is not about anything. . . . Educated Northerners like their conversation to be about ideas. . . . the notorious lack of self-consciousness of the antebellum Southerner made it almost impossible for him to define anything; least of all could he imagine the impropriety of a definition of manners" (*Col.*, 560–61). The self-expression of rhetorical talk might reach to self-realization but not to self-awareness. Nor does it have the goal-directedness fostered by linear, spatialized typographic thought. And goal-directedness can reinforce a certain kind of self-consciousness: of the self as an entity moving from here to there, among alternatives or oppositions, reaching a conclusion. Tate, evidently combative since youth (acknowledged a *calidus juventa*, of the early poem), seemingly hyperliterate since puberty, and always self-conscious, would, like Socrates, prevail at the end if he could. He distances himself from antebellum—and later—southerners by defining manners in an ironically distancing way. The celebrated charmer often in addresses and essays observes the code of manners—but as often as not in ironic fashion, as someone between rhetoric and dialectic, marking their limits.

He professes puzzlement at the end of "A Southern Mode of the Imagination" as to "what brought about the shift from rhetoric to dialectic" in southern letters, even as he posits a "New England dialectic . . . in which the inner struggle is resolved in an idea [in contrast with] Southern dramatic dialectic . . . as in . . . Faulkner, in action" (*Col.*, 568). The southern dialectic form, which he saw finally as the genius of the literary renaissance of the South, had historically to await "perception of the ironic 'other possible case' which is essential to the dramatic dialectic of the arts of fiction" (*Col.*, 568). He spent forty adult years as a man of letters never out of touch with his native and still profoundly oral South. In his later years, his health was poor, and his satisfactions in family were great. Still, one wonders whether his growing awareness of other possible cases, even ironic personal cases, was not a greater factor in crowding his public pen and voice into the intimate and dwindling quasi silence of letters.[23] But here he is one of the "Southern rural types . . . like all other people, ultimately mysterious" (*Mem.*, 148).

23. I gratefully acknowledge the discussion on these matters of my colleague J. Bainard Cowan, especially on metaphoric land.

4

WHAT CAPITAL FOR THE
REPUBLIC OF LETTERS? 🌹

I don't like the gift of prophecy. But what can one do?

        —Allen Tate, *Memoirs*

In the original unity of the first thing lies the secondary cause of all things, with the germ of their inevitable annihilation.

        —Edgar Allan Poe, *Eureka*

To talk about centers, whether logocentric or Party-centric, and about centralizing or about dispersion and decentering—as Tate and many talk—is to refer to authority. Authority is for Tate at the heart of his prophetic anxiety and fervor in behalf of himself, his South, and his America. Attention to the existential hotness of oral culture, its voice-filled worlds of contention, and its compensatory conservatisms, including tendencies toward heavy structuring in institutions, in characterization, and in mores such as southern manners reflect, affords sight of a certain irony—both important in itself, and symptomatic—in Tate's championing of culturally central structures that he, Cleanth Brooks, Warren, Ransom, and Davidson adverted to by the term *tradition,* though not quite with identical meanings. It remains to look more searchingly at his indictment of the modern enemies of "tradition" and at the prophetic, as distinct from ironic, nature of that indictment. Likewise, it remains to consider more thoroughly his recognition that modern American life increasingly requires alternatives to the prevailing structures of authority—by which I mean, as Tate does, structures of authority *felt* to be inevitable (the feel that oral culture characteristically imparts), and not to be onerous, because rights and obligations present themselves as intertwinedly *natural.*

Land, it seems, or equivalents to the land were problematical. Tate's essay in *Who Owns America?*—subtitled *A New Declaration of Independence*—has received far less attention than his contribution to *I'll Take My Stand* six years earlier but rewards attention in this context. Called "Notes on Liberty and Property," it begins by making a now-familiar argument against corporate and capitalistic giantism, drawing heavily on Adolf Berle and Gardner Means's book of 1933, *The Modern Corporation and Private Property*. Tate summarizes the first half of his essay "historically": "Since about 1760 in Great Britain and since the Civil War in America, *one attribute of property as it existed for five hundred years has been steadily lost.* That attribute is the responsibility of personal control. The other attribute remains: legal ownership. But without control its future security must necessarily be tenuous."[1] From deep in the Great Depression, he nevertheless writes, with regard to individuals as well as corporations, not of rights but of the responsibility of personal control and yet connection with community. I'll return to this. He continues, seemingly in rejoinder to the merry critics who accused the Southern Agrarians of being nostalgic ninnies who wanted every family on a small farm:

> I am not suggesting that the American Telephone and Telegraph Company break up into jealous units, one for each county. But I do suggest, if the institution of property, corporate or private, is to survive at all, that we keep only enough centralization—of production as well as control—to prevent gross economic losses and the sudden demoralization of large classes of workers. . . . Political economy is the study of human welfare.
>
> We have tried to produce as much wealth as possible. It cannot be denied that technology and corporate ownership have combined to increase staggeringly the aggregate wealth of modern States. But it is an equivocal wealth. The aggregate wealth of a nation may be stupendous, and the people remain impoverished.
>
> Ownership *and* control are property. Ownership without control is slavery because control without ownership is tyranny. . . . Pure exchange-value represents the power of its owner over other persons. Pure use-value represents the owner's liberty not to exercise power over other persons, and his independence of their power over him. The property State stands for a reasonable adjustment of these extremes.
>
> The liberty of power is the only kind possible in the corporate system. But

1. Allen Tate, "Notes on Liberty and Property," in *Who Owns America? A New Declaration of Independence,* ed. Herbert Agar and Allen Tate (Boston, 1936), 87.

liberty in the true sense is grossly caricatured when it is replaced by the mere possibility of power over our fellow men.[2]

Tate's essayistic exploration for dialectics other than use value/exchange value, and richer than power/independence, and for equivalents not only to land but, "in the true sense," to land power and land liberty—that exploration is a kind of hermeneutic pilgrimage from middle to end and on to a "new declaration of independence."

About it, I will be arguing an alternative to the immensely provocative formulation by Lewis Simpson in "The Southern Republic of Letters and *I'll Take My Stand.*" Where—though not *exactly* where—I have invoked the notions of orality and textuality, Simpson has proposed those of society and mind. He argues that Tate led the Agrarians in "enlisting mind against itself in reversal of" the Encyclopedists, whom he saw as pioneers of the Great Critique. The Great Critique is a transference of "all that human consciousness comprehends as existence into itself"—and the *all* means science, history, God, and the rest—by a third estate of clerks that Simpson echoes Pierre Bayle and Tate in calling the republic of letters. In short (too short, since the whole essay should be examined), the Great Critique is something of a totalizing myth, however often masquerading as a procedure or project, and Tate among others wished to make the South into a truer countermyth, a "symbol of a recovered society of myth and tradition," in an effort Simpson regards as intrinsically tragic. But while averring, with reference to *Huck Finn,* that "society as created by mind is intrinsically a slave society . . . it enslaves everybody," Simpson acknowledges that "each of the primary Agrarians resisted this closure [to "contemplation" or one might better say to mystery and transcendence], but Tate . . . resisted it most of all."[3] So he did, as I will argue. Yet to denominate the vastly complex and difficult-to-characterize project of the Great Critique as going from society to mind rather than from orality to literacy—though that too is admittedly inadequate—approaches a trap that Simpson appears to fall into: the confusion of mind with ideology in the sense of a sacralized system of belief with closure or totalization. Of sac-

2. *Ibid.,* 91, 93.
3. Lewis Simpson, "The Southern Republic of Letters and *I'll Take My Stand,*" in *A Band of Prophets,* ed. William C. Havard and Walter Sullivan (Baton Rouge, 1982), 70–72, 86–89, 83, 89.

ralized secularism there has been a surfeit, as Stephen McKnight has argued.[4] But one can see the Great Critique not as—at least not *necessarily* as—some kind of apostasy or Gnosticism but rather as a development in cultural history coincident with the enculturation of printing, and with rich potentialities for both good and ill. Seen that way, one's diagnosis will be less roundly of tragedy, one's prognosis not so dark as Simpson's.

Restlessly, almost obsessively, Tate's criticism of literature and twentieth-century American life oscillates from center to center, and from conception of good or bad center to conception of good or bad decentered state, not without ambivalent suspicion. Of "our cousin Mr. Poe"'s "grandiose formula for cosmic cataclysm," cited from *Eureka* as the second epigraph of this chapter, Tate goes on to say: "I read it at fourteen. . . . It lives for me as no later experiences of ideas lives, because it was the first I had. The 'proposition' that Poe undertook to demonstrate has come back to me at intervals in the past thirty-six years with such unpredictable force that now I face it with mingled resignation and dismay. I can write it without looking it up" (*Col.,* 456). Does the restless movement deserve to be called dialectical? Does it make of a fifteen-year series of essays a sort of poem, an implicit tragic drama of the spirit setting keel to the breakers of modern American life? or a hermeneutic journey, with wandering steps and slow? The chief focus here will be the essays Tate arranged and published as *Reason in Madness* (1941), *The Hovering Fly* (1949), and *The Forlorn Demon* (1953), although other essays, as well as some letters and poems, may be seen to bear on the matter. My concern will be two perhaps complementary aspects of the same thing: the stance of our cousin, Mr. Tate, and his prophetic address to some contemporary American dilemmas.

In one stereotype, Tate began as a Fugitive in Nashville, and continued as a fugitive from northern industrialism, abstraction, and positivism with a hankering for an Egyptian bondage to the lost center of a southern agrarian tradition. He had written in "Religion and the Old

4. Steven A. McKnight, *Sacralizing the Secular: The Renaissance Origins of Modernity* (Baton Rouge, 1989).

South" that "abstraction is the death of religion no less than the death
of everything else." But he never ignored, still less excused, the bondage
of black man to white, or of any man to superstition. He literally and
figuratively fled both to and from Nashville, to and from the South, to
and from the North. We see in his ambivalence to oral as well as to textual culture that there is much more to Tate than the featherless bi-
pedalism of the stereotype.

Paul Bové has recently attacked the agrarianism of Ransom, Tate,
and Warren, and simplistic views of it, in an important article, "Agriculture and Academe: America's Southern Question."[5] In addition, this article contains the best treatment at hand of Tate's early biography of
Stonewall Jackson, indicating that Tate conceived Jackson as enacting
an "image of himself" (p. 191), as a hero beyond the hero's own telling
or acting. ("Colonel Tate," indeed, but not exclusively on agrarian or
Marxist grounds.) I will return to these matters in the next chapter.
Bové's provocative critique of agrarianism, which does not always distinguish adequately between Tate and Ransom, relies on the Gramscian
idea that the dominance of the ruling class can win the "active consent"
of the ruled, that is, on the idea of a "noncoercive hegemonic apparatus." But if that idea is not a mystification, it must refer to myth. Only
a myth, a schema of reality misperceived as totalizing or at least framing, is a "noncoercive hegemonic apparatus," and I am in the present
study addressing three men's efforts to uncover and demystify just such
a mythic apparatus. Of course Tate's agrarianism would "produce no
new men in the social order, but only in imagination." That is a prerequisite to producing them in the social order. It can be "*no matter* if that
conservatism is sometimes the ground for a critique of the extended
American State" (my emphasis) only if no critique but Gramsci's signifies. And it is too early to say that Tate's critique "of the extended
State was thus inscribed within its institutions" (I make no brief for
Ransom), unless the extended State is defined without argument as all-
inclusive.[6] Here, as elsewhere in the essay, Bové seems close to a conspir-
acy theory of history. I do not deny the fact of occasional conspiratorial

5. Paul Bové, "Agriculture and Academe: America's Southern Question," *Boundary
2*, XVI (1988), 169–95.

6. *Ibid.*, 174–75, 181, 172, 193. George Core believes the usual critics of New Criti-
cism "seldom bothered to read anything but *Understanding Poetry, The Well Wrought
Urn,* and a few anthologized essays" ("Agrarianism, Criticism, and the Academy," in
A Band of Prophets, ed. Havard and Sullivan, 135). That is certainly not true of Bové,

groups, but the notion of large-group, large-scale connivance quasi-mechanically causing, sufficiently causing, large cultural deformations smacks of the very myth of the machine that Mumford, Tate, and Eiseley set themselves to dismantle.

There *is* a certain constancy in Tate's regard throughout his writing career for a conjunction of matter and spirit, of material mediums and values.[7] That is elaborated in "What Is a Traditional Society?" which appears in *Reason in Madness* but was delivered five years earlier, as the 1936 Phi Beta Kappa address at the University of Virginia. He honors their own "Mr. Jefferson" for a "special tradition of realism" in holding the "belief that the way of life and the livelihood of man must be the same." He goes on to say that "just as canvas is the medium of a painter," so traditional property in land was once the primary medium through which man expressed his moral nature, and—in a passage already noticed—that "our task is to restore it or to get its equivalent today" (*Rea.*, 218, 229). For Tate, imaginative and textualized language was to become that equivalent—but problematically traditional—medium. But for how many in his audience, or readership, or scarcely reading fellow citizenry might that be so? He himself knew language to be tricky, and imagination and textuality as well—a point for later attention.[8] He knew that a centering combination of medium and program might be unattainable or a delusion. He avers, for just one example, the "motives of all our Middle West" to be "in general, disorder attempting to correct itself by means of the further disorder of catchwords and slogans. There is no reason to infer, from the distress into which the lack of an American myth betrays us, that it is possible to create one. It is not even desirable that such a myth be created." The year is 1929, the occasion for the words above is "American Poetry Since 1920," his overview

---

whose informed and trenchant arguments reward the careful consideration I hope I have given them.

7. R. K. Meiners provocatively asserts that "the chief issue has remained the unity of being." But he construes an opposition between the mythic-poetic and the "abstractions and operational concepts of the sciences" which may fit Tate in the 1920s and early 1930s, whereas I am concerned with his deepening sense precisely of myth, in a sense of false religion, in operational thought. See Meiners' *The Last Alternative: A Study of the Works of Allen Tate* (Denver, 1963), 34, 39. And remember that *frisson* with Poe's remark: unity of being/unity of catastrophe?

8. Ann Waldon notes that "Allen *talked* Agrarian, but Caroline *was* Agrarian" (*Close Connections: Caroline Gordon and the Southern Renaissance* [New York, 1988], 101, 170). My point: he became a farmer of the dialect of high textuality.

for *The Bookman* (*Rev.*, 87). He goes on, acknowledging "Mr. Waldo Frank," to endorse the "formation of groups" as the "only effective procedure in the present crisis. . . . Only a return to the provinces, to the small, self-contained centres of life, will put the all-destroying abstraction, America, safely to rest."

Insofar as this goes beyond an advocacy of discussion groups,[9] it seems to be a prescription for poets to find a parochial centering in historical regions—not mere geographical provinces—of the republic of letters. What he recommends are multiple alternatives to "the rootless character of contemporary life . . . the tenuous substance . . . no fixed points in the firmament, no settled ideas of conscience" (*Rev.*, 88). One wonders if he was not thinking of Dante's rootless lovers, like Paolo and Francesca, eternally blown about. His prescription was probably written in New York; certainly the sometime Nashville poet here writes of poets as third persons, inviting the question of whom he is addressing and what their relation might be to those others, the poets, now that the days are evil and time needs redeeming. But even here he says that it is the job of poets to use ideas, not create them—a hint of the argument he later makes insistently that poets serve culture by representing the significance of life as it is, any life. Perhaps he is writing to a divided self of poet and critic external to his fancied own center of life.

Tate becomes very succinct indeed by the time he praises Poe as the discoverer of "our great subject, the disintegration of personality," and styles Poe a "transitional figure because he . . . kept [his exploration of that subject] in a language that had developed in a tradition of unity and order" (*Col.*, 439).[10] Insofar as the poet presents the feel of life in a horizon of disintegrated personalities, a poetics of the decentered world may confront us. Or does such a horizon imply an absent or displaced center? And what of Tate's practice in essays?

The essay on Poe was written about 1950, and the point I have cited from it differs from many earlier observations of Tate's mainly in its convenient condensation. One might note, although Tate does not quite do so explicitly, that Poe's style as Tate describes it stands in ironic relation

---

9. Louise Cowan documents discussion groups that surely endured as models for all the Fugitives, however variously influential: at Ransom's prep school, at Vanderbilt, at the Hirsches' in Nashville (*The Fugitive Group: A Literary History* [Baton Rouge, 1959], 9, 10, 17–20, 35). For Mark Van Doren's comment on the seminality of the groups for "vital poetry," see p. 236.

10. From Tate's "The Angelic Imagination," *Kenyon Review*, XIV (1952), 455–75.

to Poe's world, Tate's, and ours. Readers of Tate may all too easily fall into supposing Tate a constant partisan of a tradition, or even "the tradition," in a sense of command "by a single intentionality or privileged object."[11] But consider a relatively early remark from "Modern Poets and Convention" (1936): "Poetry in the great tradition never has more than unimportant resemblances to the poetry preceding it [and] the ordinary cultivated reader is usually the enemy of tradition; he wants to see only what he has already seen in the past" (*Col.*, 543). If this is tradition as centering, it is so in a distinctly ambiguous way. It would seem to be tradition as intentionality from *multiple* points toward a *transcendent* center, which is to say, an absent center. Tate grouped this brief piece with others as "A Miscellany" in *The Forlorn Demon* (1952), where it stands in suggestive relation with certain remarks in the preface he provided to the volume. By remarking that the divinization of the machine is widespread and an "idolatry that is in one way or another the subject of most of these essays," he suggests resistance to a false center. He elaborates that hell is "perhaps" a condition that "promptly *adjusts* and *integrates* its willing victims into a standardized monotony." Therefore, for Tate, "the modern man of letters, if he is not a play-boy, is an eccentric: he is 'off center,' away from his fellow citizen who is sure that he is standing in the Middle." Is he writing in a satiric spirit of *we, the few, are centered; they, the many, wrongly think themselves so?* No, or at least not *here,* where he concludes with a self-deprecating irony increasingly characteristic of his later years, "I sit, in doubt between waking and sleeping, on the keel of a capsized boat . . . pretending that the hull is a continent."

Tate's preoccupation with centering, because of his shifts over time and, I think, his persisting ambivalence, impinges variously on his views about the crucial matter of abstraction and of the attendant mechanization and empty or tychastic futurism of the modern world. In "Modern Poets and Convention," he goes on to state his own sense of the matter

---

11. The phrase is Joseph Riddel's, in "Decentering the Image: the 'Project' of American Poetics?" in *Textual Strategies: Perspectives in Post-Structuralist Criticism,* ed. Josué Harari (Ithaca, N.Y., 1979), 354. Riddel's whole powerful essay has helped me a great deal in pursuit of "the elusive Mr. Tate"—as another colleague has called him. Daniel Singal has rightly pictured Tate as ambivalent about the South and "rootedness." See his valuable *The War Within: From Victorian to Modernist Thought in the South, 1919–1945* (Chapel Hill, N.C., 1982). My quarrel with Singal's parsings of Tate's conflict, and its cost (disability?) follows.

as a conviction that "a poetic convention lives only as language; for language is the embodiment of our experience in words" (*Col.*, 546). Not to be embodied seems to be, by clear implication, lifeless. Loss or removal of embodiment is that abstraction which we know he considered "the death of religion no less than the death of everything else." It may be that he gradually became more negative toward the unembodied, or at least toward the secular unembodied. By 1951, he would stigmatize as *angelic* "that human imagination . . . which tries to disintegrate or to circumvent the image in the illusory pursuit of essence" (*Col.*, 413). He gratefully acknowledges Jacques Maritain's *The Dream of Descartes* (1944) for the "doctrine of angelism" and counterposes the angelic imagination to the title faculty of his essay "The Symbolic Imagination," a faculty that brings "together various meanings at a single moment . . . but the line of *action* must be unmistakable . . . temporal sequence" (*Col.*, 412). Variousness and temporality entail some decentering? He places the essay on the symbolic imagination ahead of "Modern Poets and Convention" in *The Forlorn Demon*, as if to reinforce the essay that comes later. On the one hand, he sees the drive to immediacy tending toward a temporal taxonomy, pseudoimmanence, and mechanistic or other idolatry; on the other hand, he sees a more or less mediatorial agency of word and verbal image (to be elaborated upon below, pp. 179–87) temporalizing and decentering belief and behavior.

It appears that in the late forties and at the beginning of the fifties, he was commending an intellect aware of its limitations and accordingly effective in the phenomenal world, and that he conceived the phenomenal world as for better and worse a temporalized world of culture. In "The Angelic Imagination," the counterpart essay in *The Forlorn Demon* to "The Symbolic Imagination," he asserts that "the human intelligence cannot reach God as essence; only God as analogy. Analogy to what? Plainly analogy to the natural world; for there is nothing in the intellect that has not previously reached it through the senses" (*Col.*, 453). In his splendid essay "Longinus" (1948) in *The Hovering Fly*, palimpsestic notes of nostalgia and new cheer come through. He writes of the "Greek Cosmos, an ordering of solid objects under a physics of motion" and remarks that "our multiverse has increasingly, since the seventeenth century, consisted of unstable objects dissolving into energy; and there has been no limit to the extension of analogy" (*Hov.*, 99).

Tate's entry into the Roman Catholic church in 1950 is a matter of record. Surely consonant with his attitude toward the world of experience was Catholicism's incarnationism and its liturgical emphasis—on sacrament and *leitourgos,* the people's work. Pre–Vatican II American Catholicism's strongly Thomist strain, honoring intellect even while insisting on its limitations in a fallen world, also accorded well with his outlook. But concurrently the embodiedness on which he had insisted became increasingly ironic.

His own body deteriorated. He seems never to have enjoyed above-average, or perhaps even average, strength; and his respiratory system, vulnerable from early years, was weakened by decades of heavy smoking. I hesitate between understating and overstating the point. Tate's very last poems, written in full expectation of imminent death, are touching expressions of delight in his young children by his third wife. The point is that long-abuilding structures of conviction could powerfully reinforce factors of health and personal circumstances, all concurring to help produce the near-silence of the public Tate during the last quarter century of his life.

In the earlier part of that period (1951–1968) he taught at the University of Minnesota, a public activity. And he continued his always extensive correspondence, with Warren, Brooks, Lytle, and Davidson. John Peale Bishop was gone, and there was diminishing correspondence with Yvor Winters; Edmund Wilson's unsympathetic response to Tate's conversion, cooled an always conflicted relationship. But an extensive and substantial correspondence with Kenneth Burke and Peter Taylor and Willard Thorpe and William Wimsatt and Walter Ong—the last two new in those years of the fifties and after—shows that Tate was writing privately with some energy. The correspondence with Ong discloses a plan, Tate's apparently, to organize a loose network of Catholic intellectuals for, presumably, mutual encouragement and the greater weal of the republic of letters. Resources evidently fell short for Colonel Tate's last campaign; at least there was no Catholic Intellectual *I'll Take My Stand;* his diminishing resources were focused upon maintaining his life in his divided genealogical family,[12] in his church, and in the extended quasi family of his epistolary group.

Those were labors perhaps narrower and arguably less ambitious than his book and essay projects and even his reviewing in earlier dec-

---

12. That had always been complex and frequently difficult. See Waldron, *Close Connections;* Veronica Makowsky, *Caroline Gordon: A Biography* (New York, 1989); and Walter Sullivan, *Allen Tate: A Recollection* (Baton Rouge, 1989).

ades. But he worked the language as devotedly as any good farmer might have worked the land. Writing was the labor of bodily self, and for all the cerebral quality so conspicuous in his essays and poems, he continually reminds us of the body. It is not, of course, a matter of large-muscle exertion to tote that barge; it is a matter of *voice* and interchange, and sensory reference. Many essays he delivered as talks. But nearly all read *as if* he composed them as talks. There are locutions in the vein of "I stand before you" or "I would say." There are innumerable ceremonious references to Mr. X or Mr. Y, as if in parliamentary deliberation. There are, less decisively, innumerable bits of syntax that to me read as colloquial, as speech. He occasionally refers to himself writing or to having just read something, thereby foregrounding some physical presence of self or of books in use.

What all this means is that Tate, the hyperliterate and even Euro-classicist man of letters, in the end redeems the primary and secondary orality of his southern heritage as a dynamic and ostensive thematic counter to abstraction. It is the witness that he is himself, designedly, the embodied imagination, not the aspirant to angelic imagination. He is inevitably centered in his own body, no more than, but no less than, everyone else.

That is one shadowy, uneven curve, among others, in Tate's criticism. It omits much. For one instance, it omits his relation to place, place in the geographical and physical-cultural, as opposed to the historical, sense. Place in that sense, as opposed to, say, some more dominantly rhetorical-psychological sense, has less to do with Tate as a prophet-critic than it does with Eiseley or Mumford or many another twentieth-century figure, despite Tate's often-noted southernness. Place may bear on centering and decentering, about which more in a moment. One can think of John Donne, whom Tate admired, in casting about for similarly reduced consequentiality of place: for the earlier figure as much as for the later, the close-in human situation occludes some of the immediate physicality of bedroom and church (and "countries, towns, courts"), and the historical situations of the personae somewhat occlude the middle distances surrounding Mediterranean picnic or graveyard ruminations or the physical place of writing or of speaking. We get bodily immediacy and a cultural republic extended in time, *hence implicitly shared by individual minds*—or at least shareable.

More positively, I have omitted or elided some instructive indications of substations on the American itinerary of more or less sacred discontent that are perennially habitable irrespective of the date or dates when Tate sojourned in each. Of these, failure came early in Tate's life, and it has a markedly, though not uniquely, idiosyncratic relation to centering. In a famous letter to Bishop, of June, 1931, he wrote:

> The older I get the more I realize that I set out about ten years ago to live a life of failure, to imitate, in my own life, the history of my people. For it was only in this fashion, considering the circumstances, that I could completely identify myself with them. . . . We all have an instinct—if we are artists particularly—to live at the center of some way of life and to be borne up by its innermost significance. The significance of the Southern way of life, in my time, is failure. . . . What else is there for me but a complete acceptance of the idea of failure? There is no other "culture" that I can enter into. (*JPB*, 34)

The "circumstances," we may suppose, included the irreversibility of time's arrow (he could not be defeated as a Confederate colonel), Snopesery, the suspiciously northern industrial New South, and the "homicidal varieties of wowserism" (in Mencken's phrase; Tate as a boy witnessed a lynching). Not quite "ten years ago," in 1922, the nine months Tate had spent at Valle Crucis, North Carolina, to overcome incipient tuberculosis were apparently a time for very serious thinking about vocation and poetry.[13] Whatever the realization he expressed in 1931 owes to his reflections of 1922, he was indicating in his letter to Bishop a quasi centering on a hypostatized South, but at an almost metaphysical level of analogy ironically suggestive of abstraction. It was a *quasi* centering because he recognized the ontological equivalence of other cultures better than the more oral and tribal among his compatriots did, such as his friend Davidson.[14]

But why set out, as if ironizing a fairy tale, to make one's fortune as

13. See the account by Radcliffe Squires in *Allen Tate: A Literary Biography* (New York, 1971), 36–40.

14. Edmund Wilson, in a letter of May 20, 1929, to him had pronounced him "now distinguished from many of your fellow Southerners and from most Americans of all kinds by your curiosity about and interest in the whole country and in the world in general" (*Letters on Literature and Politics, 1912–1972* [New York, 1977]).

failure? Eric Berne has argued that the idea of the "defense mechanism" gives an inadequate explanation, since withdrawal is sufficient for mere defense. In order to understand the lure of failure, it is thus necessary to ask, "What *reward?*"[15] Tate could hardly have avoided identifying among "my people" one nearest at hand who was an economic failure, his father.[16] Some men fail in order to rebuke their fathers; others both to mock and emulate them. *Failure,* too, may be a term largely reflecting a judgment by the adversary, in this case perhaps foremost the northern philistine industrialists (and is there a likeness of brother Ben's face amid those Syndics?); hence to fail may ironically be to succeed on a higher plane. To set out to fail may also be a way of commanding one's own destiny in circumstances so problematical that command is impossible to achieve in other terms. But whatever the perennial options may be for others, Tate could not endure having the South or his genealogical family as a failure that was the center for his life. Both had to be decentered or redefined—in the case of lost wives, supplanted—or else failure itself had to be abandoned as primary intentionality.[17] Although it is difficult to conceive a reliable survey of the point, unless as parody, is it not likely that most twentieth-century Americans aspire more to transcend than to emulate their familial histories? Their optimism, verging on utopianism, is part of the occasion for prophecy.

Indeed, Tate came to detest—is the word too strong?—the way of being or bearing oneself in the world that such terms as *living a life of failure, emptiness,* and by implication, *ennui* signify. He wrote to Bishop that he mistrusted the "conscience about capitalism": "I don't doubt its motives; I merely think there's something else to explain the scruple—maybe a personal emptiness. I feel that too, but with such intensity that I can't do anything about it. . . . That, by the way may be a quality I share with Baudelaire: he took his punishment and didn't pretend it was Society. I agree with him that somebody is to blame, and his name is Satan, or Evil. It is so simple that few of our contemporaries can believe it" (*JPB,* January 11, 1938). In identifying *un semblable, un*

15. Eric Berne, *Games People Play* (New York, 1964), Preface.

16. So Squires (*Allen Tate,* 21–22), but he seems to view Tate's relation to his father differently.

17. Squires sees him "shifting his thematic concern from 'failure' toward 'vision' in the 'Seasons of the Soul' [1942–1943]" (*Allen Tate and His Work: Critical Evaluations,* ed. Radcliffe Squires [Minneapolis, 1972], 7). Singal sees a shift toward a Modernist "*process* of artistic creativity rather than . . . moral beliefs" (*The War Within,* 253).

frère, and concluding the bit pugnaciously, he implies that he does have ways of doing something "about it." And he seems to acknowledge something of the sort to Bishop later: "I write better when I am in trouble, but the trouble of the world answers the purpose" (*JPB,* December 4, 1942).

One can always smash things that are felt as a reproach to one's emptiness or ineptitude. The tribe known as the Vandals gave us the common noun for that. The prophet, by generic definition, smashes delusions and idols. The more detached observer may call him an iconoclast, a breaker of the images that their worshipers believe to partake of divinity but the holiness of which others might question.

The prophet is far more iconoclastic than theophanic. Tate, accordingly, rarely announces the new name of God but rather attacks: "Yet, although Mr. [Max] Eastman is aggressive, sensational, and personal in his attacks, he has been widely read [in *The Literary Mind*]; while Mr. [Cleanth] Brooks, who is sober, restrained, and critical [with *Modern Poetry and the Tradition*] will win one reader for Mr. Eastman's fifty. Mr. Eastman . . . like the toothpaste manufacturer . . . offers his product in the name of science" (*Col.,* 117). If some have called the New Criticism as represented so notably by Brooks, Warren, Tate, and Ransom something of a museum piece, it is at the very least one much on display. And others would call it the reigning orthodoxy even yet in innumerable classrooms and many English departments. Mr. Eastman is not much on display.

Or again, more generally, Tate, like Mumford, predicted that World War II would destroy free inquiry and criticism. In "The Present Function of Criticism," he writes of the "totalitarian society that is coming in the next few years": "The actuality of tyranny we shall enthusiastically greet as the development of democracy, for the ringing of the democratic bell will make our political glands flow as freely for dictatorship, as, hitherto, for monopoly capitalism" (*Reas.,* 5–7). That is not certainly incorrect in the yet longer run, and certainly not paranoid, yet it emits a *fumet* of excess cordite.

Most grand of all, albeit from a warrior happier or at least less sardonic, is his "Religion and the Old South" for *I'll Take My Stand.* In that text of 1930 he considers religion primarily in the ecclesiastical and even denominational sense, but the religion of wealth and the ecclesia of aristocracy appear in his writing elsewhere early and late—and subject to violence: "I begin an essay on religion with almost no humility at all;

that is to say, I begin it in a spirit of irreligion. One must think for oneself—a responsibility intolerable to the religious mind, whose proper business is to prepare the mysteries for others. . . . A discussion of religion is an act of violence" (*Reac.,* 167). One should not overlook in this the weirdly deflating effect of *almost,* and the multiedged quality of the term *religion.* What Tate does not quite say is that iconoclasts tend willy-nilly to be zealots of a new religion. Thinking "for oneself," as Tate's disdain for partisans of mere process—social engineering and the like—suggests, cannot be mere process. One thinks some*thing,* and one thinks *it* in a structure or paradigm.

It is somewhat as Edward Said asserts of Vico's notion of man as a "historical creation": "The fact of his existence asserts the beginning as transgression." Or as Joseph Riddel observes of Charles Olson: "The man who is the *first* to look *again* at the 'start of human motion' repeats the Original Act, the First Murder, in the sense that he fractures the dream of some original, recuperable purpose." [18] Whether Tate intended the like or not, it comes to that.

And nothing much less sweeping than that can fit some of Tate's pronouncements about poetry that began early and have puzzled readers. In 1932, for example, he unobtrusively implicated himself with other poets, including an arresting pair:

> Personal revelation of the kind that Donne and Miss Dickinson strove for, in the effort to understand their relation to the world, is a feature of all great poetry; it is probably the motive for writing. It is the effort of the individual to live apart from a cultural tradition that no longer sustains him. . . . The poet in the true sense "criticizes" his tradition, either as such, or indirectly by comparing it with something that is about to replace it; he does what the root-meaning of the verb implies—he *discerns* its real elements and thus establishes its value, by putting it to the test of experience. (*Reac.,* 18–19)

The "test of experience" means honoring the concreteness of place and personal history but not totalizing.

18. Edward Said, *Beginnings: Intention and Method* (Baltimore, 1975), 353; Riddel, "Decentering the Image," 349. Olson's phrase is from "Song 4," in *The Maximus Poems* (Berkeley, Calif., 1983), 19.

During the thirties he made a nostalgic suggestion that became more equivocal later: that the authentic core of a decayed tradition can be renewed. The spirit of this idea is analogous to that of the English reformers, including Donne, who wished to revive "primitive" churchmanship. A nostalgia for mythic—and lost or spoiled—origins apparently persists indefinitely as one option for Americans; it drew Tate while the depression years darkened. Still, the critical stance, living a little apart and putting things "to the test," is characteristic of all languages in contrast to wordless infancy, and also of literacy in contrast to orality. Tate republished his essay of 1936 "What Is a Traditional Society?" five years later, with its extreme language and apotheosized Jefferson: "Man has never achieved a perfect unity of his moral nature and his economics; yet he has never failed quite so dismally in that greatest of all human tasks as he is failing now. . . . To behave morally all the time [at ease and work—] this principle . . . is the center of the philosophy of Jefferson" (*Reas.*, 228–29). *Never failed so? Never!* This is prophetic hyperbole, not squarely opposed by the principle at the center of Jefferson's philosophy, both because of a likely incongruity between *center,* as concept and metaphor, and *unity,* with reference to what is compound, and because Tate claims the centering principle only for Jefferson's philosophy, not his practice. Tate's predominant attitude toward tradition is evident when he writes of looking forward to Bishop's essay on the New York World's Fair: "It will be the nearest I'll get to that graveyard of an era" (*JPB,* May 24, 1939).

To replace a tradition that "no longer sustains him," Tate looks no more than Yeats to a *worldly* utopian vision. In the "Suppressed Preface" that he finally published in 1946, he insists rather, "I have tried not to know too much about what I wanted the world to be" (*Hov.,* 102). It is an earned self-description, and it captures Tate's distaste for making idols. So, too, does the exception he sometimes felt he had to take with fellow poets and critics, some of whom were friends: "I agree with Mr. Ransom that Mr. Blackmur's position is untenable. . . . I hope that Blackmur will not try to hold a position and that he will continue to be interested in something else; for example how the imagination acts in a given instance" (*Col.,* 539). The same is true of his writing of "the period between the wars: the period, in fact, in which Mr. MacLeish and Mr. [Van Wyck] Brooks said that we had staggered into a war that might have been prevented had the men of letters not given us such a grim view of modern man from their ivory towers, or simply refused to be concerned about him" (*Col.*, 402). The "position" and the idealized or exemplary view not to be held—opposite the scapegoated "view of modern man"—are idols or idolatrous, it seems, for the Tatian life world. Particularist acts of the imagination or views of them and of modern life, even when such views of poetry and modern life may look expansively toward "modern man," are decenterings.

So too can history be either an idol or a decentering: "The history must prove itself in the poetry, not the poetry in the history" (*Reas.*, 156). The contention occurs in the context of a review essay on David Daiches in 1941. Tate's generic taking-seriously of Daiches' work is reinforced by his judicious tone and by expressions of respect. But all that is framed by the polemical metaphor of Tate's title: "Procrustes and the Poets." It is a proper question whether poetry itself thereupon becomes a kind of idol, a quintessential partner to small-farm in the land-or-letters equivalence. Not, I think, in the essays of *circa* 1935–1955 that are our principal focus here. A slightly earlier remark to Bishop can helpfully summarize the Agrarian element and define it by partial contrast: "Our [Agrarian] point of view, which, of course is briefly: the dole must be capitalized and its beneficiaries instead of remaining idle must be returned to the land. The big basic industries must be broken up and socialized and the small businesses returned to the people" (*JPB*, April 7, 1933).

The clauses about basic industries having to be broken up and small business *further* decentralized, and the awkwardness of treating so pervasive a condition as land as a privileged object, all reinforce the point urged earlier, that even in the Agrarian "point of view" there is a diffusion that seems incipiently decentering. Seven years later, in "Hardy's Philosophic Metaphors," Tate points to relationships of land and language in terms that apply about as well to himself as to Thomas Hardy: "His 'advanced position' is only another way of saying that he had very early come to be both inside and outside his background, which was to be the material of his art: an ambivalent point of view that, in its infinite variations from any formula that we may state for it, is at the center of the ironic consciousness" (*Reas.*, 122). "Background," one readily understands, is intended here for Hardy, as it clearly is for, say, "Clym Yeobright," as Wessex land. Land—with its culture—and language are intermixedly rather than disjunctively "material of his art." An "ambivalent point" of view, at the "center" of consciousness—like a consciousness whose irony by the nature of consciousness pervades the subject's life world—implies an idoloclastic decentering. Is not an ironic consciousness precisely one with competing candidates for center? It challenges and disrupts "any formula."

Tate uses the notion of center ironically: "The mind is the dark center from which one may see coming the darkness gathering outside us" ("Preface," in *Reas.*, ix). The mind, that is, may be a reflector or reciprocal or point of hermeneutic dialogue but *not* the maker of the world. He stigmatizes as provincialism, that is, as temporally constricted ethnocentrism, our failures to discount the "nineteenth century dream of a secular Utopia" (1945; *Col.*, 287) and today's "widespread eschatology of a . . . naturalistic Utopia of mindless hygiene and Tom Swift's gadgets."

The comment on the eschatology of hygiene and gadgets was written at the beginning of the war, and the paragraph concludes in words that could be Mumford's were it not for the characteristically Tatian invitation to irony implicit in asserting, "There is no doubt": "There is no doubt that the most powerful attraction offered us by the totalitarian political philosophies is the promise of irresponsible perfection in the future, to be gained at the slight cost of our present consent to extinguish our moral natures in a group mind" (*Reas.*, 103–104).

Many critics have attended to Tate's corresponding point, that the historical imagination is an "exercise of the myth-making propensity" and that "under positivism we get . . . historical method . . . the way of discovering historical 'truths' that are true in some other world than that inhabited by the historian and his fellow men" (1936; *Reas.*, 222–23). Although everyone who is familiar with many of the essays sees that Tate's ideological bête noire was likely to be termed "positivism," I have not seen anyone assimilate that passage from "What Is a Traditional Society?" or the point in general, to a rationale of opposition to idolatry and to its intimately associated pseudotranscendences ("some other world"). That opposition can regard tradition and positivism, past and future, with equally chilly eyes.

Similarly, in "The Function of the Critical Quarterly," Tate provocatively asserts, "If the reader is not encouraged in self-knowledge—a kind of knowing that entails insight into one's relation to a moral and social order that one has begun, after great labor, to understand—then taste and judgment have no center, and are mere words" (*Col.*, 66).[19]

19. Originally in *Southern Review*, I (1936).

"Insight into a center," antithetical to "mere words," still betokens some metaphysical, or at least taxonomic, center. In contrast, "knowing . . . one has begun" betokens dynamic process. By the time of his essay "Longinus," he is counterposing the Greek cosmos and "our multiverse" (*Hov.*, 90).[20] In "Is Literary Criticism Possible?" he shifts the emphasis: "By rhetoric I mean the study and the use of the figurative language of experience as the discipline by means of which men govern their relations with one another in the light of truth. Rhetoric presupposes the study of two prior disciplines, grammar and logic" (*Col.*, 476).[21] Of course, *light* witnesses to the visual emphasis preeminent in a post-Gutenberg, textualized world. But Tate posits it here as a condition of apprehension, not as its sufficient cause, nor particularly as a center. And "use . . . discipline . . . govern[ing] their relations"—in a manner seemingly beyond the taxonomies of grammar and logic—such language looks toward the Augustinian journey and the "task of the civilized intelligence" as "one of perpetual salvage" ("Ezra Pound and the Bollingen Prize," 1949; *Col.*, 536).[22] The "truth," above and outside time, by the light of which men might formulate in words constructive interrelationships and freely social action, seems to have become, more and more clearly for Tate, love (see Chapter 5). On February 9, 1955, he wrote to Lytle: "I come more and more to the view that much of our agrarian thinking twenty years ago was much more modern than we realized: we were setting up the means as the end. . . . Having agreed that the agrarian is superior to the industrial society as a means, we tended to rest the case there, and thought of the agrarian as the end."

The truly prophetic mode, toppling idols and smashing icons defined by the prophets as false, leads to decentering. Or else it honors too privileged an object, too single-minded an intentionality about what the world should be, and decays into further idolatrous mythmaking. The point has been elusive for at least three reasons. First, Tate's *professedly* nonreligious position, followed by a widely acclaimed entry into Ca-

20. Originally in *Hudson Review*, I (1948).
21. Originally in *Partisan Review*, XIX (1952).
22. Originally in *Partisan Review*, XVI (1949).

tholicism in 1950, has misled some, Edmund Wilson among them, into seeing him as an apostate from something like humanism, even though he never accepted that attribution; others, mirror images of the first group, have seen him as Saul of Tarsus, Tennessee, turning into Paul of Damascus, Minnesota. Second, many of Tate's associates and many of the critics who have written about him came of age during the long generation—say, the 1930s through the 1950s—when *mythic* and *mythmaking* and *mythographer* were honorific terms. At least they were widely used as honorific terms because they could point to an activity of the humane imagination honorably resisting one or another reductively rationalistic fractionation, abstraction, or mechanization in thought. Having—dare we say?—metabolized the gains from that long campaign, many of us would now beware of mythologizing as, if not inevitably, at least in usual practice, baneful mystification, politically, ideologically, personally. Without wishing to style Tate our Roland Barthes any more than our Saint Paul, I conclude that he represented extensively, cogently, and for many years a quasi-Hebraic, prophetic, antimythic position in opposition to idols of totalization. "A higher unity of truth" we may believe in "even if [one supposes a rhetorical concession, supposes he believed "even though"] it must remain beyond our powers of understanding" (1950; *Col.*, 482). Third, Tate himself seems intermittently to represent a centralizing position, and one surely congenial to some of his readers. Consider:

> There should be a living center of action and judgment, such as we find in the great religions. ("Humanism and Naturalism," 1929; *Reac.*, 139)

> Social man is living, without religion, morality, or art (without the high form that concentrates all three in an organic whole) in a mere system of money references through which neither artist nor plutocrat can perform as an entire person. ("The Profession of Letters," 1935; *Col.*, 276)

> Both politics and the arts must derive their power from a common center of energy. (Preface, 1936; *Reac.*, x)

> The disintegrated, multiple "psyche" of modern man is intelligible only with reference to the historic unity of the soul. ("Liberalism and Tradition," 1936; *Reas.*, 203)

> *All* that [Dante] knew came under a philosophy which was at once dramatic myth, a body of truths, and a comprehensive view of life. ("Understanding Modern Poetry," 1940; *Reas.*, 83)

The logical opposite, or the historic complement, of the isolated community or region is not the world community or world region. . . . The complement . . . is a non-political or supra-political culture such as held Europe together for six hundred years and kept war to the "limited objective." ("The New Provincialism," 1945; *Col.*, 285)[23]

How can this anthology of centrism be squared with the earlier-cited advocacy of decentering? It is at least more useful for this generation and indeed more correct to treat decentering as Tate's primary concern. But it would be sloppy beyond reasonable bounds of interpretive freedom and uncertainty to ignore the reiterated centrist note. At least three factors prompted reiterations of that note.

First, Tate from hot youth, *calidus juventa*, to elegiac old age was the very opposite of the cool rationalist superciliously regarding the world from a sidewalk café (as Gabriel Marcel characterized Jean Paul Sartre). He was a lover, a detester, and profoundly a man of family. Like other writers uneasy in parental and marital family (in Tate's case, his first),[24] he invested himself heavily in an extended quasi family of longtime intellectual companions. And some among them dearest to him—Lytle, Warren, Davidson, Cleanth Brooks, Ransom, John Bishop, Peter Taylor—were "traditionists" (their word), in ways less complexly ambiguous than his own way. Tate was an independent and principled man, but his capacity for deference should not be overlooked, nor was he unaffected by southern manners.

23. The anthology could be enlarged with citations a little narrower, or a little less explicit. See, for example, *Reas.*, 20; *Hov.*, 72; *Col.*, 283. See also letters to Davidson, especially the early and Eliotic ones of June 23, 1925 ("emotions . . . *eccentric* [versus restoration of] poetry to the tradition . . . an integration of sensibility"), and May 14, 1926 ("culture is dissolving"). Early, more than later, one can indeed find what Paul Douglass styles in Tate, Ransom, and Warren a "full flowering of the Modernist organicism whose aesthetical roots lie in Bergson's *Creative Evolution* and William James's *Principles of Psychology* (1890)" (*Bergson, Eliot, and American Literature* [Lexington, Ky., 1986], Chapter 5).

24. See Tate's *Memoirs and Opinions*, the published correspondence, Squires's *Allen Tate*, and above, 111*n*5. His unpublished correspondence with his daughter is as eloquent as F. Scott Fitzgerald's roughly analogous letters to his daughter. The somewhat marginal but real connection of household turmoil to my line of argument is suggested by Ford Madox Ford's letter to Dale Warren in June, 1937 (after some weeks with Allen and Caroline): "Consorting with the Tates is like living with intellectual desperadoes in the Sargoza Sea. We have just concluded an attack upon the University of Vanderbilt" (*Letters*, ed. Richard M. Ludwig [Princeton, 1965]).

Second, he was variously uneasy or at odds or exasperated with some allies and friends not so traditionist as himself. The published reviews, references, and correspondence (primarily letters to Bishop) indicate this with regard to Wilson, MacLeish, and Winters, perhaps to Ransom, more ambiguously to Burke. Those indications are confirmed in the unpublished correspondence. He was often at odds with Wilson.[25] Ransom always discarded letters, so we have surely lost a fascinating adjunct to Tate's respectful demurrers to Ransom's rationalism. It seems doubtful that Tate's contributions survive from his forty-year correspondence with Winters. Winters is wonderfully contentious, and one infers Tate was little less so. The Tate-Burke correspondence resists brief comment and indeed merits separate publication and analysis. But the general point holds that Tate, though an independent man, could be moved to counterattacks conveniently mounted *from* some central tradition *against* what he perceived as attacks from ostensibly rationalist directions.

Finally, there is the inconstant heart—and the more embattled or the wearier, the more likely it is to be inconstant, at least inconstant in the sense of feeling nostalgic moments of longing for what Auden once named the "warm nude ages of ancestral poise." In 1949, Tate exchanged a series of letters with Karl Shapiro over the award of the Bollingen Prize to Ezra Pound and his own vote for that award. Deeply troubled at what seemed to him his friend's misconstruals of his position, he wrote, finally: "I think as I've thought for years that the 'liberal' point of view ends in fascism or communism; the traditional restoration, or recreation, is just about our only defense against those twin tyrannies. At the bottom of the liberal misconception of tradition is the mistaken notion that fascism was reactionary: it was a nihilistic revolution."[26] The strenuous disavowals of tyrannies of right or left, of liberalism, the impromptu if not beleaguered quality of *just about*, the enormous ambiguity of *restoration, or recreation,* and the anxious care to be transparent to his friend in the republic of letters perhaps sufficiently

25. Wilson wrote Tate on February 15, 1955, about getting Pound out of Saint Elizabeth's Hospital. Tate, on the back, in pencil: "This, as usual, is completely wrong-headed" (Tate Collection, Firestone Library, Princeton University; published with permission of Princeton University Library and Helen H. Tate).

26. Allen Tate to Karl Shapiro, April 27, 1949, in Manuscript Division, Library of Congress; quoted by permission of Helen Tate.

bring together the three factors behind Tate's occasional centrist note and return us to his descrials of what to pursue if not a center.

He characterizes ironically the spatial binary centering of polemical Us *versus* Them: "Rhetoric in the Reconstruction South was a good way of quarreling with the Yankees, who were to blame for everything" ("Faulkner's 'Sanctuary' and the Southern Myth," 1968; *Mem.*, 146).[27] Provincialism means to him a kind of parochialism in *time,* and he can with a more rueful irony acknowledge that a situation in which every moment seems a new origin, that is, a situation of decenteredness in time, is vulnerable to a kind of myth of the moment: "Provincialism is the state of mind in which regional men lose their origins in the past and its continuity into the present, and begin every day as if there had been no yesterday. We are committed to this state of mind. We are so deeply involved in it (I make no exception of myself) that we must participate in its better purposes, however incomplete they may be; for good-will . . . is better than ill-will" ("The New Provincialism," in *Hov.,* 31). He refers to Christopher Dawson and acknowledges with appreciation perhaps too scanty in irony a "peculiar balance of Greek culture and Christian otherworldliness, both imposed by Rome upon the northern barbarians." But *balance, imposed?* "Is not this civilization just about gone?" ("The New Provincialism," in *Hov.,* 26–27).

Perhaps it is, or perhaps *gone* is a quibble, a prejudicial interpretation of *change.* Much earlier, in "A Note on Donne" (1932), he sketched in a preliminary way an alternative he returns to, an alternative for life in a postclassical, post-Christian, neobarbarian present. He writes of a "poet's ideas . . . pitted against one another like characters in a play" and remarks, "Therein lies the nature of the 'conceit' . . . an idea not inherent in the subject, but exactly parallel to it, elaborated . . . into a supporting structure. . . . The conceit in itself is neither true nor false. From this practice it is but a step to Dryden and the 18th century, to the rise of the historical consciousness, and to ourselves" (*Reac.,* 72).[28]

27. On rival brothers and symmetrical conflict, see René Girard, *Violence and the Sacred,* trans. Patricia Gregory (Baltimore, 1977).

28. A "function of a critical quarterly," Tate said in a *Southern Review* essay of that title in 1936, is "to discredit the inferior ideas of the age by exposing them to the criticism of superior ideas"—partly drama, partly military engagement, partly perhaps the greater light subsuming the less. The combination is at least a sign of his many-mindedness, and an inconsistent metaphor if implicit.

Later, in "Miss Emily and the Bibliographer" (1940), he envisions a dialectical (his word) cultural activity transcending polarities of imaginative writers and literary historians. He stigmatizes the literary biographer, and any such before him in the Princeton English Club that April day, as worse than Faulkner's Miss Emily, who pretended that something dead was living, because, with literature, they pretend that "something living is dead." He concludes:

> We must judge the past and keep it alive by being alive ourselves; and that is to say that we must judge the past not with a method or an abstract hierarchy but with the present, or with as much of the present as our poets have succeeded in elevating to the objectivity of form. For it is through the formed, objective experience of our own time that we must approach the past; and then by means of a critical mastery of our own formed experience we may test the presence and the value of form in works of the past. This critical activity is reciprocal and simultaneous. (*Reas.*, 114–15)

The animus against any mythic, centric *method* or *hierarchy* is recognizable enough, and one might argue, as I will, that Tate elsewhere gives persuasive meaning to the notion of "formed, objective experience." But what of the enabling "critical mastery of our own formed experience"? One possible implication—a significant one deserving more attention than it has had in the educationist establishment—is that each of us should write autobiography.[29] To live a successful American life in the age of mechanism requires the subversive act of writing one's life.

More immediately, he comes to speak admiringly of Longinus' "profound but topical dialectic" (*Hov.*, 89) of *techna* (art, skilled making) and *hupsa* or *ekstasis* (joy, transport). We have lately heard from cultural critics, psychoanalytic, anthropological, and theological, a good deal about the serious value of *absorbed play*; see, for example, Bruno

29. Not that it has not had some movingly serious attention. See, for example, works by Sylvia Ashton-Warner and James Herndon, with regard to primary and secondary education. At the college and graduate level, interest in autobiography grows, fostered by the scholarship of James Olney, William Spengemann, and others. Not the least provocative argument in Lewis Simpson's powerful "The Critics Who Made Us: Allen Tate" (*Sewanee Review*, XCIV [1986], 471–85) is that the essays collected as *The Man of Letters in the Modern World* (beginning with the title essay and ending with "Narcissus as Narcissus," from 1937) enact the "drama of the man of letters" as the "crisis of recognition that distinguishes the tragic mode" (p. 484). I recognize the self-dramatization, but perhaps stubbornly see Tate both in fact and in his own perception locating the tragedy mainly in his society's embrace of baneful myth.

Bettelheim, D. W. Winnicott, Gregory Bateson, and Harvey Cox. But Tate has pioneered and remains almost uniquely prophetic in proposing a pleasure of the text, a kind of dialectical *jouissance,* which would be a "declaration of independence from the practical, forensic eloquence of the rhetoricians" (*Hov.,* 90). *Forensic eloquence* we can recognize here as an idiosyncratically metonymic term for fetishistically purposeful or goal-centered discourse, by those who would honor power to the eclipse or even exclusion of love. The practical and the forensic that he disdains are defective because too narrowly self-interested, because deficient in that love called "sense of obligation."

So, by the Tate of this interpretation, we are both literally and symbolically called not to poetry as greater than history but to poems (*Reas.,* 156), called both ideologically and ostensively to irony inside and outside our traditions and situations (*Reas.,* 122, and *passim*) lest we collapse into idolatry or Gnosticism, called from any place collapsing into self to an ethereal but dynamic network of intimacy, called into time extended by formalized, dialectical oscillation of present and past lest inchoate time collapse into a prison of mere present moments, called to an institutional landscape furnished with literal and also, I think, with symbolic "critical quarterlies," each imposing one or another "concentration of purpose" to "order . . . a scattering experience" (*Reas.,* 193). It was Tate's oscillation away from mythic and tyrannous totalizing toward liberating openness and pluralism, away from annihilating disjection toward concreteness of place and history and focal love and *pietas—this* great dialectic that gave a peculiar American point to what was no mere temperamental edginess and to what were premonitory warnings in a more than generational time of troubles.

5

THE REPUBLIC OF LETTERS: HOMESTEADING AND BEDSIDE

Benjamin's Berlin book is proof of the constitutive role of distance [and] is devoted not so much to memory itself as to one of its special gifts, which is captured in a sentence from his One-Way Street: *"Like ultraviolet rays, memory points out to everyone in the book of life writing which, invisibly, glossed the text as prophecy." . . . "Certain words or pauses allow us to detect the presence of that invisible stranger, the future, who left them behind with us." . . . Benjamin wrote no more city portraits in the period after 1933. . . . With the loss of one's homeland the notion of distance also disappears. If everything is foreign, then that tension between distance and nearness from which the city portraits draw their life cannot exist.*

—Peter Szondi, "Walter Benjamin's 'City Portraits'"

Where is there nevertheless a reference? And this reference seems to me not alongside other possible objects, but has to do with the whole of our world view and of our exposure to ourselves in life. "Being" is not an object—to reformulate Heidegger.

—Hans-Georg Gadamer, "The Eminent Text and Its Truth"

With the epigraphs, I mean, as usual, to cast a sharp, raking light on the realm of discourse to follow. Any possible extended analogy—of Tate as our Benjamin, in the manner that Izaak Walton urged upon his countrymen that Donne was their Augustine—I leave to others to draw. But in Chapter 3 something became evident of that "tension between distance and nearness" Tate felt in and out of the South, and something of its analogue in the *relative* distance of literacy and writing as against the voice-filled and animated oral world. In Chapter 4, against the background of the initial considerations, I foregrounded Tate's restless canvassing of what he called the "crossing of the ways" in Southern history—which he judged analogous to that in Renaissance English cultural history—as well as his reading of prophecy in that book of life and his restless rejection of any mere object "alongside other possible objects" as a privileged center or transcendent being.

Here I want to foreground the Tate mainly of some later essays, the Tate not of warriorship for all seasons but rather of a poignantly vulnerable triumph of letters. First to be considered is the potential civic triumph of letters, a limited triumph of truth that might make men free—without necessarily making them happy—if they have eyes to see and ears to hear; second, the private triumph of letters as the potential medium of what Gabriel Marcel might have been glossing Tate in calling our presence to ourselves; finally, the ambiguous if not indeed mysterious "letters" that constitute the medium, the substance, even the structure, of the only republic he very greatly believed in.

Tate wrote in 1950 that "the human condition must be faced and embodied in language before man in any age can envisage the possibility of action" ("To Whom Is the Poet Responsible?" in *Col.*, 405). He regularly directs us to notice men acting mindlessly or mechanically or in animal-like fashion—being defaced. But the seeing *face* of self-identity, and possibility as a seen face, partly a recognizable future self *envisaged*, are more fundamental for him. He is even more a lover than a hater, and the speaking subject implicit for him potentially in the face of any situation or text belies at least in privileged instances the mechanical deadness of print, the tomb of text, and the pastness of occasion, however occasional all texts may be. In "Our Cousin, Mr. Poe," he writes about Poe in order "to set forth . . . a little of what one reader finds in him, and to acknowledge in his works the presence of an incentive . . . to self-knowledge. . . . His voice is so near that I recoil a little" (*Col.*, 470).

There may even be an element of the willful about it, in the sense of the consciously arbitrary. It is to be kept in mind that in "Miss Emily and the Bibliographer," he writes that "it is better to pretend with Miss

Emily that something dead is living than to pretend with the bibliographer that something living is dead" (*Reas.*, 102). The bibliographer is *no less* than Miss Emily a pretender; Faulkner's text is somehow alive. The preferable pretense seems to contribute in Augustinian fashion to the reign of love (*ad regnem caritatis*); Tate might have judged the bibliographer's to be, in Rebecca West's contemporaneous phrase, "plainly damned."

In 1952, he explicitly rejected as illusion the notion he ascribes to the "secularists" of the modern city, "Baudelaire's *fourmillante Cité*," the notion that "they have no hell," an illusion to which they are vulnerable because they have "lacked the language to report it" ("The Man of Letters in the Modern World," in *Col.*, 384). Tate, like Mumford and Eiseley, articulates language to report hell and death, giving them, through his *report*, a kind of paradoxical life, at the same time that he seeks to make present the lively—albeit in a construction quasi-detached from life by textuality.

Written and printed language may afford a purchase on the reality of urbanized culture corresponding to what the agrarian smallholder's land gave upon earlier culture, if the two are the mediums of analogous labors of loving stewardship. Two of Tate's essays anatomize, with a sharpness he never surpassed, the presence to ourselves and others of the critical intelligence of ourselves and others, unavoidably attended by the mediation and thus distancing of writing, the presence of what he uneasily calls "*imagination*" (1943) and "love" (1952).

The later piece, "The Man of Letters in the Modern World," has a generality that situates and glosses the brilliantly particular, and cryptic, "causerie" that he originally called "Dostoevsky's Hovering Fly." He begins with the familiar differentiation, rendered as a subordination, between being and doing. The man of letters *is* what he *does:* he does "recreate for his age the image of man, and . . . propagate standards by which other men may test that image, and distinguish the false from the true" (*Col.*, 379). Since it appears more at large that the man of letters is a potential realization of any person not disabled by ignorance, derangement, or myths of mechanism, this amounts to a momentous definition: man is man doing, man from, man for, man toward. Part of the doing, as Tate construes it, is distinguishing a grammatical and a logical function as subordinate elements of language, thought, and action. To do so in language may be as dynamic and decisive as voting with one's feet. But never in lockstep; the goal is to instruct "a sufficient minority of persons," for human participation is precisely noncoercive, mediated presence, what Tate had come to mean by *rhetoric.*

Differentiating the false and true, or the worse from the better, is to battle against threatening effacements of difference. I accept René Girard's argument that loss of differentiation, or threatened loss, is cultural crisis.[1] It may be for Tate, although this is less surely the case, a death equivalent. He moves to suggest the antithetical figure, the presence of the man of dark letters. In what I take to be a stratagem to disarm or defamiliarize, he calls it a "digression" that will "not condescend to Descartes by trying to be fair to him." In one of the most Manichaean moments in his career, he poses Descartes, notwithstanding some disclaimers, as the adversarial "strategist of our own phase of the war" of man with himself. "What happens in one mind may happen as influence or coincidence in another," and Descartes has "isolated thought from man's total being, . . . divided man against himself," and "mechanized nature." Tate will not easily fall into a determinism more appropriate to the adversary: "influence *or* coincidence." He moves to disarm a certain objection, acknowledging an "almost mythical exaggeration" in the reproach he voices. But his "almost" quasi-epically leaves the opposition quasi-epically intact. Yet, his "digression" has moved to historicize his differentiation between historical man, almost Kristevan man *en procès,* and ahistorical, mechanized man who functions or else is repaired or discarded.

He writes in part for the University of Minnesota audience he would speak his writing to, an audience he seems to anticipate responding dialogically even if mostly in the tacit dimension of communication.[2] He writes of the "melancholy *portrait* of the man who stands before you" (my emphasis). He would suppose that audience, and *a fortiori* the later readership, to be only fractionally of the Roman Catholicism he embraced, would suppose it to be largely Judeo-Christian, but more casually so than himself in many instances. To them, Tate may be understood to relate as in tertiary parody to so conventional a message as that in a familiar hymn: "Wake, awake for night is flying; the watchmen on the heights are crying." The hymn is a sacred parody of the *aubade,* of

1. René Girard, *Violence and the Sacred,* trans. Patrick Gregory (Baltimore, 1977), esp. Chapter 1.

2. I mean to acknowledge, though not to explore here, the relevance not only of Mikhail Bakhtin's principle of dialogism but of Michael Polanyi's *Tacit Dimension* (Garden City, N.Y., 1966).

profane lovers parted by the threat of discovery; Tate is concerned with a range of love between the erotic and the advent good news of divine love. A reiterative pattern of his imagery warns of deepening darkness if we fail to wake and light the lamp.

But he addresses more directly and polemically the large part of any academic audience who, often without having thought much about it lately, take themselves not to be religious. In effect he argues, Not likely. Tate, who cites Maritain and Reinhold Niebuhr, clearly parallels Paul Tillich in taking the object of one's ultimate trust as one's actual religion. Hence, he wryly attributes to—or projects upon?—society a sort of Jansenist "intractable Manicheeism" (*Col.*, 381) with regard to itself, that is, a tendency to hypostatize its darkness or badness. Accordingly, too, one large modern (and perennial) group can be observed to idolatrize means, and they constitute a religion in "Europe eastward of Berlin, and in Asia" even more, he judges, than in the West, where their presence and strength engender crisis (*Col.*, 382–83).

If we have difficulty thinking of "society as the City of Augustine and Dante," we may—and Tate implies we will—come, like the communist societies, "to prefer the senility (which resembles the adolescence) and the irresponsibility, of the barbarous condition of man." He surely knew Milton's scornful lines about preferring "bondage with ease, to strenuous liberty." He probably has in mind the Freudian idea of defense mechanisms when he muses ironically, "There is perhaps no anodyne for the pains of civilization but savagery" (*Col.*, 383). Did he know parts of *The Renewal of Life?* The good city as analogy to adult individual maturity and to medieval society suggests intimate commonality—the web of obligation.

In the same place in another historicizing gesture he quotes Samuel Johnson on drunkenness as anodyne. Tate may not have known the social magnitude of the alcohol problem that was brought about by cheap gin in eighteenth-century London or that is besetting twentieth-century Western societies, but he was sufficiently aware how problematical his own and his first wife's drinking were that Johnson looks like a stalking horse. In any case, he verges on Eric Berne's neo-Freudian argument that explanations of actions in terms of *defense* must be supplemented with attention to the *rewards* provided by the lines of action one is trying to explain. Worship of means, like inebriation, may reward with a sense of partial detachment or achievement and may reward by occluding from consciousness some painful conflicts implicit in results, but not indefinitely. Reality—it bears repeating—is more arrogant than any scheme or tactic.

Tate goes on to say that "what men may *get out of this*" (my emphasis), this pseudorewarding worship of means, is "in the western world today, an intolerable psychic crisis expressing itself as a political crisis." He declines in the following paragraph to assign decisive priority to the psychic or the political—and naturally so. Who has not observed both the person deranged by political crisis and the person who materializes politically—in family, office, or state—the turmoil he feels internally that seeks its reflection at large? Tate renews the Renaissance interfiguration of self and state.

As with Mumford and Eiseley, a highly literate, writerly sense of time fosters Tate's most resonant sense of specifically human doing. That is, he puts aside the mechanically foreshortened temporality and ahistorical automation of stimulus and response, interest compoundings, and the like, for a much larger, less clear, but more significant structure: "Man is a creature that in the long run has got to believe in order to know, and to know in order to do. For doing without knowing is machine behavior, illiberal and servile routine, the secularism with which man's specific destiny has no connection" (*Col.,* 383). Moreover, "generation after generation . . . throughout history," doing without knowing, doing in a context of information rather than humanely capacious knowledge, of "communication" rather than "communion," has been power without love and is plainly damned.

The parallel with West's words from the epilogue to *The Meaning of Treason* might seem forced if Tate were not so explicitly talking about hell, and similarly sorting out the true and false imagers of men. Sadly unaware of Gabriel Marcel, like most in American letters at the time, he writes of "our contemporary existential philosophy, a modernized Dark Night of Sense," of "Baudelaire's *fourmillante Cité*," of the man of letters and "his hell." More pointedly, he asserts the "barbarous disability" of the banker and the statesman insofar as they live in the aforementioned "illusion that they have no hell because, as secularists, they have lacked the language to report it" (*Col.,* 384). It seems that words do not make hell, but without mediative *utterance* in words, hell has them (see below, pp. 175–79).

Experience, for Tate, is not necessarily linguistic, but a humane mastery of experience is, and it will include the vocabulary of history, of the "long run." That long run is part of the truth of being human which transcends any machine truth and is the glory of history as burden. Perhaps in keeping with Tate's indirectly avowed "intractable Manicheeism," his sense of the burden of history alternates with his sense that sin, as in the Thomist notion, is a privation of being, *peccatum nihil*. The modern corporate hell is in serious part a kind of miserable privation of language for truly identifying itself—like Milton's Mammon proposing to spruce up hell: "What can Heaven show more?"

The likeness is no casual one. Tate, like his honored poetic forebears of the seventeenth century, strives to resolve or decently evade the dilemmas of historic orthodoxies concerning what they normally call sin and (worldly) hellishness. It is easy to reject a facile doctrine of progress, that sin and hellishness are mere error or ignorance by assiduity correctible. It is easy to reject by appeal to experience something of the doctrine that *peccatum* is *nihil* yet that "intractable Manicheeism" is a heretical distress to orthodox faith. It is easy to eschew the language of original sin and, as a post-Romantic champion of imagination, advocate discernment and choice. Yet when were sin and hellishness not *there* in human life? When was *choice* not so problematical and implicated in darkness as to rebuke Pelagianism?[3]

Tate's image as if from fairy tales or parable—of a hell to be named if it is to be escaped—may strike us as more somber than Donne's or Milton's symbolic wayfarers on the stony ground of history, more somber even than West's throngs of historical refugees. Tate accords the matter of naming, direction, and chosen orientation similar centrality in resolving the complex of issues. But he draws on that other great Western metaphor of being: *theatrum mundi*. In his version, we are thrown into the playhouse in midplay and *must write the true and lively script*. His frame of urbane exposition and judicious polemic may mask some of the violence and darkly allegorical force of his words. There is, he says,

> evidence, generation after generation, that man will never be completely or permanently enslaved. He will rebel, as he is rebelling now, in a shocking variety of "existential" disorders, all over the world. If his *human* nature as such cannot participate in the action of society, he will not capitulate to it, if that action is inhuman: he will turn in upon himself, with the common gesture which throughout history has vindicated the rhetoric of liberty: "Give

---

3. The distinctions in this paragraph owe much to Paul Ricoeur's essay on original sin in *The Conflict of Interpretations: Essays in Hermeneutics* (Evanston, Ill., 1974).

me liberty or give me death." Man may destroy himself but he will not at last tolerate anything less than his full human condition.

Would it be fair to object that Tate has ignored or sentimentalized those all too numerous who have been making this an egregious age of state-sponsored torture, those who have capitulated fully to the inhuman action of society? The objection would be inaccurate, surely. He seems to have considered the likes of the torturers and their masters to be already dead, by *their* gesture. In denying the fellow presence of the Other, they have absented themselves from human life. In any case, he addresses not the authors of inhuman action in society but the situation of those subject to it or those who are accomplices by bewilderment or passivity.

Exactly, then, as an act of consequential connection, of mediative presence, it is the "business of the man of letters to call attention to whatever he is able to see . . . to render the image of man as he is in his own time, which, without the man of letters, would not otherwise be known." The man of letters is precisely the prophet, not the angelic imagination, because concretely immersed in his own time and culture. Still, his prophecy exceeds the reportorial and representational and offers renderings of what would "not otherwise be known." He is to make the images known to some moiety potentially decisive. The images will be neither the secret property of a Gnostic priesthood nor signifiers orbiting in a realm of other signifiers. Tate's appeal is to other minds, however language-articulated or text-driven those might be.

Social actions and social rhetorics that one gesture may participate in, another gesture may occlude or twist or mock. The point is made by a roll call of images that figure assorted rejections of societies of idolatrized means, gestures assortedly costly to self yet made in behalf of "our full human condition." They range from Emma Bovary and Hepzibah Pyncheon to Stephen Dedalus and Joe Christmas. All literary and, accordingly, different from ourselves, separate and absent, they nevertheless are called "our ancestors and our brothers" and are implicitly as present to the literate consciousness as our Aunt Polly or Grandfather William.

About halfway through the twenty-four-paragraph essay, in paragraphs 12, 13, and 14, Tate resumes explicit attention to the notion of communication. He returns to ironize, defamiliarize, and differentiate. Is the "problem of communication" solved by the mass media? No, communication cannot occur merely "by means of sound over either wire or air." Complete communication, beyond something we merely "*use*" (Tate's emphasis), involves communion through love. "We *partic-ipate* in communion"—*in:* neither a realist nor an idealist nor even a phenomenological mere surface is in question here.

Is communion useless? Presumably not, because it will have a proxi-mate occasion and intentionality (away from isolation, say, or conflict), and an ultimate intentionality of return from fallen status to God, a step in the mind's ascent. Tate's Christian orthodoxy is insistent here, and he confesses it with artful irony. The "simple-minded Evangelist" is counterposed to the "otiose ear of the tradition of Poe and Mallarmé," not quite the favored port of call in the University of Minnesota or *Hud-son Review* of the 1950s. But in any case, for him, self-realization as a goal, whether conceived in Christian terms or not, transcends in the order of being mere operational goals.

But the irony reinforces the disavowal of sentimentality. He explicitly disavows "an obligation of *personal* love towards one another," indeed acknowledges any man may hate other men, and may love "his neigh-bor, as well as the man he has never seen, only through the love of God." By its counterposing of possibilities within the self against one another, or posing some joint element of self and reader against a bit of the text, the irony likewise dramatizes a Coleridgean point he has em-phasized about freedom.

"The medium," Tate quotes Coleridge, "by which spirits understand each other, is not the surrounding air; but the *freedom* which they pos-sess in common." That freedom is presumably a concomitant, in a con-sciously Augustinian sense, of loving others for God's sake lest one love them merely for one's own. Certainly it is a concomitant of a generally linguistic approach to the self and of an awareness that the particular linguistic sign has a take-it-or-leave-it, noncoercive status. In contrast, a conception of communication in the reductive mold of use-sense repre-sents the sign as having the status of an agent in something like a train of gears or levers. Coleridge's mutually possessed freedom is for Tate a me-dium of mutually respectful, or he might say loving, language.

The self-styled simpleminded evangelist, known to be southern, cere-moniously invokes the remark of "Mr. Auden," known to be neither southern nor simpleminded: "We must love one another or die." Au-den's pronouncement serves Tate's purpose, and in an Augustinian way. Not to love persons for God's sake is, he assumes, not to love anything else for God's sake, and thereby to be a proper citizen of a secular so-ciety, by which he means a "society of means without ends." If means do not have their ends somehow in God, the implicit argument runs, then as with attitudes toward persons, the only alternative is a welter selfishly reductive of human ends and desires. Tate anticipates the language of Joseph Fletcher, in *Situation Ethics,* by implying that if the ends do not justify the means, nothing else will.

But Tate makes a point more particular to any society that is, like ours, "in the age of technology": it "so multiplies the means, in the lack of anything better to do," that "our descendents will have to dig themselves out of one rubbish heap after another." But they will scarcely have the wherewithal or understanding to do so: "The surface of nature will then be literally as well as morally concealed from the eyes of men." In 1952, probably some in Tate's listening and reading audience were familiar with Mumford's decrying of the degradation of cities. But many no doubt thought a vision extending the rubbishscape to oceans and all was a bizarre hyperbole unless Tate meant nuclear apocalypse.

To what extent does the public language, that cultural equivalent, present analogous alternatives? If Tate was thinking more than glancingly in Auden's terms, he may have thought of the "farming of a verse" to "make a garden of the curse" (as in the elegy on Yeats). Clearly and surely he did drive onward toward an image not only of "man as he is" but of letters (whether or not law) like love. And although a long paragraph of concessions, beginning with the "congresses of men of letters," acknowledges there to be—almost in Auden's words in "Law like Love"—much that "like love we seldom keep," Tate's rhetorical posture is more confident and decisive than Auden's. One measure of that confidence is his short way with nostalgia: American writers are under demand "to 'communicate' quickly with the audience which Coleridge knew even in his time as 'the multitudinous Public,' shaped into personal unity by 'the magic of abstraction.'" One supposes that *magic* signified to Tate hidden manipulation, and the associations of *abstraction* we know as deathly.

He makes of his admired acquaintance Alexis St.-Leger Leger, the poet St.-John Perse, a differentiating and overdetermined symbol. At first glance, the former permanent secretary of the French foreign office might seem a reassuring figure, at least to any American "still able to think that he sees in Europe . . . a closer union, in the remains of a unified culture, between a sufficiently large public and the man of letters." But the physical distance of Europe from America, the semantic closeness of the idea of remains to that of rubbish, and the almost oxymoronic internal clash of *closer union* betoken the somber conclusion: "Two names for the two natures of the one person [*i.e.*, St. Leger and Perse] suggest the completion of the Cartesian disaster, the fissure in the human spirit of our age . . . and the eventual loss of communion."

Yet Tate proclaims the man of letters in the modern world to be somehow *in* it, rather than beside it, viewing it from a sidewalk-café table. Privately, he might have adduced the insurance executive–poet Wallace Stevens, with whom he poignantly failed to achieve communion, as a fissured spirit.[4] What of that other, more notoriously public man of letters, no friend to Tate or his poetry, whose own poetry Tate found of mixed quality but who got his vote for the Bollingen Prize? In public essay and fervent correspondence, Tate justifies his vote to give Pound that prize on the basis of Pound's service to language.[5] Surely this man—and not the only one—in some part repellent to Tate affords a useful partial perspective on what Tate takes the man of letters to be in a world he sees in terms not so much of *usura* as of Cartesian split, triumphant "positivism," and idolatrized means.

Tate seems to mean Pound's service to language as farmerlike service to the written signifier analogous to Christian stewardship. Tate regards Pound's signifieds, insofar as this poet conceived himself a statesman and economist, as intrinsically worthless. Tate confidently infers those signifieds from Pound's signifiers—which both are and are not Tate's own, or ours. For Tate, Pound's signifiers are, in their better segments and moments, situatings in definite place and time of the human communion as experience. In a review of *A Draft of XXX Cantos* for the *Nation,* he praised those cantos as not *about* anything so much as like, in Pound's own words, "conversation between intelligent men" (*Rev.,* 124–30).[6] And he commended them as not given "up to any single story or myth." In short, Tate even then was celebrating an ultraliterate intelligence, an adept of signifiers, who exemplified communal concern with history and myth but was too literate and ironic to be prey to history as determinism or myth as totalizing, though Tate was already in

4. After several exchanges of letters, Tate wrote a deeply serious and personal poet-to-poet letter. Stevens responded with chat about Christmas shopping and getting a haircut in New York. Tate's penciled comment: "If Stevens has ever done a generous thing, I never heard of it" (December 9, 1949, in Tate Collection, Firestone Library, Princeton University; published with permission of Princeton University Library and Helen H. Tate).

5. Notably with Karl Shapiro. See above, 152*n*26.

6. Originally in *Nation,* June 10, 1931.

communal fashion also eager to give Pound credit for some sense that latter-day seductive and dehistoricizing myths can turn men into swine.

One may grieve for the tragic blot on Pound's escutcheon. One can regret, even if a little unfairly, that Tate—rather like José Ortega y Gasset in Spain—did not see quite with the clarity available later that advanced acculturation to reading and writing, to the variously communal, playful, ironic, sinister manipulations of signifiers, fosters for well and ill the divergent specializings of a Chaucer, Perse, Stevens, Pound, and lesser men. Specialization is literate and measures our distance from the tribalism of primary orality wherein everyone must help to keep everything vocalized and remembered and the language and culture are almost a single tool, as Eiseley was to put it.

In any case, it is such conscious presence-with-distance, in effect such mediated presence, that Tate champions: a communal consideration of "*end, choice,* and *discrimination,*" as against communication as the infliction of stimuli upon drives to secure a response. It is, quite consistently with Tate's ideological opposition to the myth of the machine and the attendant idolatry of means, an opposition to the practice of mechanism and to the susceptibility to mechanism.

"Works of literature," accordingly, "from the short lyric to the long epic, are the recurrent discovery of the human communion *as experience,* in a definite place and at a definite time" (Tate's emphasis). They are recurrent discoveries because new works get written in each generation. They are recurrent because of new readers and readers of readers (young Keats looking into Chapman's Homer, or whomever). They are recurrent because of the rereadings afforded by the quasi fixity and semipermanence of print. Such writerly or readerly discoverings are experience in Tate's terms insofar as they involve imaginative presence through affectional sympathy or the love of God and insofar as they involve at least fractional rearticulations of the reader's existential horizon and sense of human possibility.[7]

7. Obviously the issue is omnipresent and complex with regard to Jacques Derrida, and I defer generalization. But see, for example, near the end of "Freud and the Scene of Writing," in *Writing and Difference,* trans. Alan Bass (Chicago, 1978): "the concept of substance—and thus of presence." The French original of that essay appeared in *Tel quel,* XXVI (1966); the French edition of the collection appeared in 1967. One supposes, with regret, that Tate's declining health kept him from reading and commenting on the European reconsideration of language and culture much after he read Jacques Maritain and (apparently) Albert Camus and Jean Paul Sartre. But he corresponded with Walter Ong.

Moreover, the role of the man of letters is prophetic: the master genre of literature in this age of myths and idols—of false totalizings—is prophecy, the genre of revisionary subtotalizings. It may at one moment, more in the ironic mode, foreground revision, at another moment, more in the georgic mode, foreground a subtotal. Such is the implication of discerning, mapping, and signifying "what has not been previously known about our present relation to an unchanging source of knowledge" (*Col.*, 389). The prophet of letters is the one who must neither neglect history nor fail to subordinate it to eternity and who must always critique and reinvent myths and images yet always somehow situate them in respect to the Judeo-Christian Other that by definition transcends every mythic or symbolic definition. No doubt that is one among others of Tate's reasons for privileging no image so much as the imageless doing or bearing implicit in loving for God's sake. That literature in its proper imaging of the mediums of, and our present circumstances of, caring may use—as will come out—something so humble as a hovering fly is in keeping with an incarnation in a stable. In fact, the Tatian prophet *must* ground any pieties in the language of particular time and place—time and place experienced, even if mainly in the imagination (rather than in the language of merely affected time and place, as Tate felt MacLeish had done with Mexico in *Conquistador*).

Tate, however, compounds the issue almost to perplexity. He insists on the linguistic nature of our necessary understanding but invokes nonpolar metaphors. Of the "two sets of analogies" he identifies—*drive, stimulus, response;* and *end, choice, discrimination* (*Col.*, 388–89)—the first he interprets as "sub-rational and servile, the other rational and free," and he situates his interpretation by insisting that the "analogies in which man conceives his nature at different historical moments is [*sic*] of greater significance than his political rhetoric." Thus he invites the scrutiny of a fundamental myth or metaphor that may be obscured by political or, as we might say, *presenting* metaphors, metonymies, or synecdoches. He immediately applies and exemplifies both the point about man and that about "greater significance":

When the poet is exhorted to communicate, he is being asked to speak within the orbit of an analogy that assumes that genuine communion is impossible: does not the metaphor hovering in the rear of the word communication isolate the poet before he can speak? The poet at a microphone desires to sway, affect, or otherwise influence a crowd (not a community) which is then addressed as if it were permanently over *there*—not *here*, where the poet himself would be a member of it; he is not a member, but a mere part. He stimulates his audience . . . thus elicits a response, in the context of the preconditioned "drives" ready to be released in the audience. Something may be said to have been transmitted, or *communicated;* nothing has been shared, in a new and illuminating intensity of awareness.

Orbit and *illuminating* seem to exemplify metaphors of the degree of significance of "political rhetoric." Perhaps *communicate* expresses a planetary notion implicitly revolving with other such around a solar notion of idolized means—or positivism—which is to be understood as controlling it at a distance. But Tate does not repeat the metaphor or put it to a critique. *Illuminating* seems a bow, perhaps ambivalent, just possibly careless, to the venerable association of light with wisdom and godhead. It seems in this paragraph, and often in Tate's writings, to be overborne by the sense conveyed in the word *intensity:* the sense of liveliness and intimacy. Hence illumination in his sense is missing from the existential darkness of mere communication (that is, manipulation) without communion.

More perplexing, he seems to be damning *distance* in some sense: "*there*—not *here.*" Yet the "metaphor hovering in the rear of the word" *communication* seems to be the Latin meaning of *communicare* as "to impart" or (his gloss) "to transmit"—like a disease, with like invaded by unlike across time and space. Similarly *drive* and *stimulus,* from the Old English *drifon,* akin to the Old Norse *drifa,* "to dash" (as of spray), and the Latin *stimulus,* "goad," akin to *stilus,* "stake." *End, choice,* and *discrimination* do indeed contrast with that "specifically animal mode" both in usage since the seventeenth century and in etymology. *End* apparently derives from words immemorially meaning "end" back to the Gothic, and even the Greek cognates *ante-* and *anti-* associate with a temporality or spatiality much more intellectualized and less coercive than *goad,* as in *antepenultimate* and *antithesis,* as Tate the classicist knew. So too with *choice,* related to the Old French *choisir,* the Old English *ceosan,* the Latin *gustare,* and, it seems, to the Gothic *kausjan,* "to examine," "to test." If Tate does associate *choice* with *gustare,* any discordant associations with a "sub-rational and servile" taste in the mouth that might arise for him may be dominated by the notion of that *gustibus* concerning which there is no disputing (because it is ineluctably personal). No metaphors incipient in the second set of terms seem essentially adverse to freedom.

So Tate problematizes, even if not altogether consciously, the closeness that is good and the closeness that is bad, the distance that is bad and the distance that is necessary. (Compare "Reach out and touch someone" and "Put the arm on someone.") Insofar as Tate's argumentative and metaphoric implications are more coherent than centrifugal, participation or communion seems to involve neither unity nor a common center so much as a shared degree of control and a shared degree of risk. The implicit good space is not void or geometical, nor is it *necessarily* the quasi-filled space of audition. Rather it is the intimate space-between that constitutes a mutually respectful margin of choice: not Aeneas carrying Anchises on his back just ahead of flames and destruction and ghostly injunction but rather Adam and Eve walking away from fallen paradise hand in hand, or communing reader reading. *Communication,* in Tate's sense, is less risky to the manipulator, less controlled by the manipulatee, than is needful in either case for humane life.

Insofar as Tate's assertions and implications decenter themselves centrifugally, they testify to the difficulty in gauging and finding, the more so in maintaining, that exact distance which best serves love and freedom. Augustine's Christian teacher, confronted with perplexed interpretations, is counseled to choose the one conducive to the reign of love (*ad regnem caritatis*). There is likewise with Tate, in effect, some substitution of vector for distance, in that the mappings of the literary imagination imply steps on a path.

Writing letters in either the general sense—of the man of letters—or the epistolary sense may be more humane and dignified than resort to the voice that cracks, or coerces with a "certain pitch of sound." Communion as shared and equal control and risk tends to foster something with a life of its own, as we see in friendships and projects that survive vicissitudes and in our inner life with the images "that passion, piety or affection knows." Derrida, in "Freud and the Scene of Writing," concludes that Freud was tending toward a view that "writing is *techne,* is the relation between life and death, between presentation and representation."[8] So might anyone tend, with emphasis falling now one way, now the other.

For Tate, the emphasis clearly dwells on the side of life. Whatever communion we are blessed with in loves and families, whatever the convivialities of our friendship—and Tate was strong in both those categories—whatever our pieties, Tate would have us humanize ourselves and

8. Derrida, *Writing and Difference,* 228.

our addressees by writing. It is as if he conceives that presence and absence cease to pose a dilemma or an undecidable choice if we regard human presence—in contrast, say, to a dog's way with its nose—as *essentially* three elemented, involving self, Other, and symbolic medium. As the chapter began by proposing, to conceive of those three elements is not necessarily to spatialize and taxonomize, as we do in saying, "triangular," or, "Reach out," or, "Don't make waves"—which we do because we would *see* a sign.

Tate goes on in the twentieth of his twenty-four paragraphs to direct any of us—regardless of gender, I take it, *anthropoi* of letters—to scrutinize "the letter of the poem, the letter of the politician's speech, the letter of the law; for the use of the letter is in the long run our one indispensable test of the actuality of our experience" (*Col.,* 390). Writing authenticates, and it permits discriminations by the highly literate mind, discriminations closed to orality, even to genius if merely oral. Accordingly, we can discount, although not entirely, his complaint (see above, p. 166) about fissures and specializations and dissociations in the Western mind. They are the dangerous concomitants entwined in the bright promise of textualized society.[9] Yet, almost by definition, the danger of the new (or newly recognized) thing is more *urgent* than the opportunity it presents.

So the man of letters is at once to promulgate the "difficult model of freedom" and to "*discriminate* [meaning "attack"] abuses of language, and thus of choices and ends," to attack "mechanical analogies in which the two natures of man are isolated and dehumanized," to attack "usurpations of democracy that are perpetrated in the name of democracy," to attack anything that fosters enslavement in any way, including definitions of human life as mechanistic, definitions obviously or subtly or incipiently mythic.

Having set in circulation his vocabulary of major ideas about the ominous state of the modern world and the calling of the man of letters there, having thereby exemplified that calling to a consequential portion of his listener's or reader's attention—having, that is, presented his vocational self—having begun his last paragraph with ironic analogies to "*parvenu* gods and their votaries of decaying Rome," he concludes by making explicit the most radical definitions and professions of faith implicit before:

9. See above, Chapters 3, 4.

The state is the mere operation of society, but culture is the way society lives, the material medium through which men receive the one lost truth which must be perpetually recovered: the truth of what Jacques Maritain calls the "supra-temporal destiny" of man. It is the duty of the man of letters to supervise the culture of language, to which the rest of culture is subordinate, and to warn us when our language is ceasing to forward the ends proper to man. The end of social man is communion in time through love, which is beyond time.

Tate thus extends culture in two directions beyond Mumford: toward a religious transcendence about which Mumford was ambivalent, and toward the materiality Mumford would distinguish as "civilization." Maritain's term by itself might seem to smack of unitariness verging on taxonomic stasis. But Tate's words *way, through which, forward, ends* (in the plural) not only link hands with countless biblical, patristic, liturgical, and Renaissance literary images of the pilgrimage of life, they also cohere as a significant variation: the image is of an activist existential path, almost a Heideggerian *Weg,* that fulfills itself. Caring smallholdings in the culture of language can contend in dehumanized culture in analogy with, and with only slightly less hopeful prospects than, the City of God's perfusing of "decaying Rome." Is Tate perhaps after all—I borrow Walton's trope for Donne—"our Augustine"? [10]

In the "causerie" "Dostoevsky and the Hovering Fly" (1943), subtitled "Poetry and the Actual World," and reissued as the title essay of *The Hovering Fly,* Tate speaks of the last scene of *The Idiot,* with Myshkin and Rogozhin by the dead and sheeted body of Nastasya Filippovna. The scene holds such power for Tate that he takes it to epitomize the use literature can have. We are invited, it seems, to extend his idea to the use of his own essays. What *needs* are fulfilled by reading as he does and writing as he does? How should I, without pretending to some corporate voice, say he invites us to read him? to read the Other?

He says that "we cannot imagine" the scene without Myshkin and

---

10. I suggest it with more confidence for having benefited from Robert S. Dupree's fine study (focused on Tate's poetry). *Allen Tate and the Augustinian Imagination* (Baton Rouge, 1983).

Rogozhin "unless like a modern positivist we can imagine ourselves out of our humanity; for to imagine the scene is to be there, and to be there, before the sheeted bed, is to have our own interests powerfully affected" (*Col.*, 162).[11] Perhaps we *can* start to imagine versions of the scene without Myshkin or Rogozhin. But we may quickly conclude that such versions are both different from and less than "the scene," tending toward the reportorial and even statistical: "Woman Found Dead." Tate makes the point, oddly in advance of recapping the scene, that he is not berating "the sciences, but only the positivist religion of scientists. I am even more concerned with what [the scene] leaves out. . . . First, the actual world; second Dostoevsky's hovering fly. . . . But we shall not know the actual world by looking at it; we know it by looking at the hovering fly" (*Col.*, 156).

Tate's strategy appears to be to take up his predesignated topic of poetry and the actual world in nonthreatening—perhaps even disarming or seductive—generalities and to bring himself and his audience at last to the strong medicine of Dostoevsky. To highlight Tate's implications, conscious or unconscious, I begin with his Dostoevsky. His point is both explicit and implicit. "For me, the room is filled with audible silence. The fly comes to stand in its sinister and abundant life for the privation of life, the body of the young woman on the bed," murdered by Rogozhin, "of course[,] . . . since in no other way may he possess her" (*Col.*, 157). Almost as explicitly, the buzzing prominence of this fly which will inform us of the actual world is synecdochic: thus, does art, at its most brilliantly successful, inform us of the situation of carings and failings in a hostile universe of "biological situation" (*Col.*, 159).

Yet how necessarily do we "imagine the scene [and thereby come] to be there, before the sheeted bed"? There, for Tate, "the buzz of the fly distends, both visually and metaphorically, the body of the girl into the world. Her degradation and nobility are in that image." Do we concur? We may hesitate. There is much of the visual about the scene, and we silently *read* of the buzz that is "like a hurricane" in the "audible silence."

It is too late to interrogate the scene as merely Dostoevskian. Even in my sketch of it, it is Dostoevsky and Tate's scene, with the attestation of an existential perspective in which the buzz contrasts like a hurricane

11. On history as interaction with the dead, see also José Ortega y Gasset, *Man and People,* trans. Willard Trask (New York, 1957), esp. 158.

with the silence that is "audible" because it comes after a voice is significantly stilled and the dead body imports to the merely naturalistic world nonnaturalistic horizons of degradation and grandeur. Tate's attestation, like mutely significant bones in the dry valley, or new buildings more or less in keeping with the older architecture of the city square, is no less present than the words of Dostoevsky he quotes.

The self that Tate shows, doubtless differing from his biographical selves of that and other years and doubtless removed from ourselves somewhat, yet betokens a congenial version of ourselves in its concern with apprehending and representing what the world is (else why would we be reading?) and its concern with attendant implications as to what the world needs of us ("what path we ought to be on"). "We shall not know the actual world by looking at it; we know it by looking at the hovering fly." If we do that existentially strenuous thing, the threatening muteness and taxonomic spatiality of typographic culture are redeemed as a site and sight of mediating presences. With the personal *enterprise* to make them that, there is lifted what some have deemed the curse of presence when it is construed as forever a lost paradise or a deferred one. Tate's classicism, and the classical notion of the eyes as windows of the soul, would provide a seat for the conviction that the seen textual element, rightly understood as mediative—that is, as dynamically connecting, not least ethically—is as intimately present to the inner self as anything can be. The textual sheeted bed attended by textual Myshkin, Rogozhin, and Tate will in an obvious sense tell us less of pain than the deathbed of a loved one, but it can by its very placement in the between-space of noncoercive mediation tell us more about the world, including the worlds of actual deathbeds.

The mediative scene, into which one puts oneself and one's anxieties about death, guilt, and meaninglessness,[12] and which one puts into one's psyche, will nevertheless by its very textuality tend to control threatening losses of difference or losses of connection. Tate seems in his early years to have felt some threat of engulfment by family or by the southern complex that Nashville represented. He rejoiced in the freedom of New York City, as his letters of that period attest. In the thirties, *I'll Take My Stand* and other writings, including myriad letters to southern friends, along with his returns to Tennessee, evidence an anxious determination not to lose connection with his native South. (Warren, in con-

---

12. Paul Tillich's triadic formulation in *The Courage to Be* (New Haven, 1952).

trast, seemed long to welcome such disconnection, for all the devotion of his correspondence with Tate and other southern friends.)

Tate seems to imply that the self will enjoy certain rewards if it actively construes scenes, textual or otherwise, as mediative, especially if the scenes have a degree of complexity like the one including dead Nastasya, good Myshkin, bad Rogozhin, and interpretive Tate. The self will not be oblivious or cretinously "positivist" but will rather be in touch with life in some larger sense; it will be a polyvalent self.[13] And the self's ethically consequential awareness will stand in contrast to the radical otherness of a fly-buzzing nature. The contrast will be humanely evaluative, though Tate throughout his career disclaims any Arnoldian notion that poetry might substitute for religion in a secular society. Much religion is poetic, but poetry is not necessarily religious, and no poetry is adequately religious.

Tate has in this crucial essay maneuvered his way—to repeat his own characteristically martial figure—like an army in the field, from relatively easy observations and engagements with generality to the dangerous citadel of Dostoevsky's scene. I began with the intensities of that scene, in order that they might help to disentangle Tate's views and implications. Some of his remarks leading up to the presentation of the scene can serve as a reprise, and an address to a further point: "At the catastrophe the resolution of the dramatic forces is not a statement about life. . . . The "meaning" of the actions is conveyed in a dramatic visualization so immediate and intense that it creates its own symbolism. . . . Great . . . poets . . . give us something that is coherent and moving about human life which partakes of actuality but which is not actuality as it is reported to me by my senses" (*Col.*, 146, 153).

His point is not simply the familiar one that imaginative language in prose or verse transcends the merely propositional even as by its mediative agency it forgoes some of the supposed immediacy of sensory moments (though it transcends something of the uniqueness of any one such). For one thing, his overall argument and his treatment of the silence and the hurricanelike buzz in the textual scene suggest that he suspects sensory experience to be always somewhat mediated. If my senses

---

13. Compare Martin Heidegger's recognition of the self not simply marshaled "at the ready" but implicated in the mysteries of desire and the Other ("Question Concerning Technology," in *The Question Concerning Technology, and Other Essays*, trans. William Lovitt [New York, 1977]).

report "to me," they must do so through biological and cultural mediums that cannot be inert or neutral. The sensory report may largely avoid history, the mediative report will "partake" of history, and either one will be *to* a *me* neither altogether sensory nor altogether linguistic. Reality is mediated, and understanding is hermeneutic. The point has come to be a familiar one, but not one for which the New Critics are conventionally given much credit. Yet as early as 1936, in arguing with regard to Kenneth Burke and the historical environment, he advocated decentering any "Cause—Victorian Morality or Marxian doctrine"— by the "exercise of the critical wit . . . in the whole context of experience."[14]

"Something that is coherent and moving": should we say the dual insistence drives along the ridgeline between arid formalism, including that disguised in the materiality of historicism, on the one hand, and solipsistic impressionism, on the other, and between what Susan Wells has recently and somewhat differently argued as poles of typicality and indeterminacy?[15] It is better, I think, to say between a topography of knowing and an economy of doing. Tate descries an "impulse to reality which drives us through the engrossing image to the rational knowledge of our experience" (*Col.*, 161).[16] That unnamed impulse, somewhat like an Aristotelian delight in knowing, not improbably akin to the motive of Donne's climber up the huge hill of Truth or of Milton's gatherers of fragments of the "torn body of our martyred saint," suggestive of Voegelin's urge to transcendence—that impulse "must also be one," part of what makes humanity "one from the beginning [as] 'poetry is one'" (Tate approvingly cites Socrates to Ion). And it is a oneness that is at odds with the loss of boundaries suggested by *hurricane*, or *engrossing*, yet that is not delimiting or totalizing.

Similarly, in "The Present Function of Criticism" (1940), Tate rejects David Daiches' formulation of a historical critic who would work *down* from a view of civilization to the work of art or a formalist critic who would—maybe, sometimes certainly, with loss of reality—"work up to

14. Allen Tate, "Mr. Burke and the Historical Environment," *Southern Review*, II (1936), 369.

15. Susan Wells, *The Dialectics of Representation* (Baltimore, 1985).

16. Compare Tate's essay of 1936 reprinted as "Liberalism and Tradition": "Since the most significant feature of our experience is the way we make our living, the economic basis of life is the soil out of which all the forms, good or bad, of our experience must come" (*Reas.*, 230).

civilization as a whole." Tate rejects the tendentiously metaphorical *up* and *down*, rejects the putative divorce of what he calls a "formalist monster" from history, and insists in his final sentence that "literature is the complete knowledge of man's experience, and by knowledge I mean that unique and formed intelligence of the world of which man alone is capable" (*Col.*, 14–15). "Complete knowledge" means not Gnostic totality but rather poetic language stabilized—in comparison with readers' lives—in text, whereby it can provide an effectually shareable medium of symbolic inclusiveness for readers not disabled by unloving mechanistic operationalism. By what is "unique" to man and "formed intelligence," Tate clearly means the triumphs of language, which he further glosses in "The Hovering Fly" in the generic if not indeed ontological categories of grammar, logic, and rhetoric. Tate's rejection of up and down, his insistence on yoking form, history, the world, and above all, experience, to include readers reading—all this implies the hermeneutic circle. His hermeneutic circle is to be understood not as a static emblem but more as a perning in a gyre, moved by love, whether that love be adequately self-regarding and communal, or debased and degenerate.

Without that rational knowledge of our experience, reached through the "engrossing image," our experience "is mere process"—kinesis and sensation of the least mediated kind, it seems. So sense impressions constitute something of a middle term between rational knowledge, in which they are implicated as potent contributors, and mere process, in which they would seem to be passive elements, at the ready.[17] Tate gives little attention to the problematic quality of this ambiguity, and gives consequential attention in particular only once to the more or less literate eye reading texts. He does, just there, acknowledge the claims of oneness to be an "immoderate deduction beyond any preparation he [that is, Tate] has been able to ground it in." The oneness is, in any case, not so much oneness of a center as oneness of poise, intentionality, and *via*. Tate's subsequent formal communicancy in Roman Catholicism might be taken to identify more clearly, but never altogether clearly, the terminus of the *via*. And "engrossing images" would seem to involve the (threatening?) welter of the unconscious.[18]

17. See also Denis Donoghue, "On Allen Tate," in *Allen Tate and His Work: Critical Evaluations*, ed. Radcliffe Squires (Minneapolis, 1972), esp. 289.
18. Exactly a *via*, and the associations of *engrossing* were not clearly at hand when he wrote "Literature as Knowledge" for the *Southern Review*, VI (1941). See especially

His own explicit figure, implicative of process and *via*, is a decentered but not contrifugal balance of forces. One's grammatical self, logical self, and rhetorical self, "when in proper harmony and relation, achieve a dynamic and precarious unity of experience." He defines that unity by contrast, relates it to Dostoevsky's scene in the passage about the self properly there "before the sheeted bed . . . our . . . interests powerfully affected" (see above, p. 173). "The grammarians of the modern world have allowed their specialization, the operations technique, to drive the two other arts to cover," he argues, "whence they break forth in their own furies, the one the fury of irresponsible abstraction, the other the fury of irresponsible rhetoric." Even the philosopher may be found to serve the operational technique, "whether in the laboratory or on the battlefield." Moreover, the poet—and the poet is the rhetorician, the specialist in symbol—serves the operational technique because, being the simplest mind of his trinity [Tate is relentless, not sparing himself], his instinct is to follow and to be near his fellow men." But the point is less the excess of grammar than the possibility of harmonizing out defects of the hermeneutic circle. Tate, classicist and student of the text-based arts of language that have developed in history in definable contrast to orality, takes the classical pedagogical sequence of the liberal arts as existential generic absolutes, permanent bearings or ways of being in the world. By doing so, he commits the dynamism of self to the cultural dynamism that is history—and always unpredictable.

Nor is it out of keeping with his previous identifications of the modern malady as "positivism" that he extends the figure of the grammarian to be synecdochic for the emotionally arrested technician of equipment or institution. The combative "Colonel Tate" of the thirties might not acknowledge, as the wryly self-implicated and hence ironically militant Tate here acknowledges, that "like a modern positivist we can imagine ourselves out of our humanity," can embrace the "great modern heresy," can become manically predatory or otherwise camp followers of operationalism.

But he has not gone soft. The ironist who could dryly characterize

---

Reas., 60–61. "A Reading of Keats" (*American Scholar*, XV [1946]) continues the later view, insisting criticism must be comparative: "Ode to a Nightingale" is an "emblem of . . . the impossibility of synthesizing, in the order of experience, the antinomy of the ideal and the real" (*Hov.*, 63).

the rhetorician as a simpleton (to be called tragic?) [19] following "his fellow man" even to destruction concludes by casting the coldest eye of his career on the cold eye:

> From what position shall the critic, who is convinced that the total view is no view at all, the critic not being God, and convinced too that even if (which is impossible) he sees everything, he has got to see it from somewhere . . . [,] under what tree, then[,] . . . shall the critic take his stand, which may be less than an heroic stand, to report what he sees, infers, or merely guesses? (*Col.,* 149) [20]

> The fiction that we are neither here nor there, but are only spectators who, by becoming, ourselves, objects of grammatical analysis, can arrive at some other actuality than that of process, is the great modern heresy: we can never be mere spectators, or if we can for a little time we shall probably, a few of us only, remain, until there is one man left, like a solitary carp in a pond, who has devoured all the others. (*Col.,* 162)

Mere spectators are persons deprived or self-deprived of mystery and hence of a need for hermeneutic circling. "Process" is either flux, meaningless to grammatical taxonomies, or else it may hermeneutically become literature and history, dynamic in its presence, present through sympathetic imagination and the vital stability of language crafted with care for medium and user. Tate's essay is more than a Lukacsian fragmentary form of longing, more than any fragmentary and half-dramatic plea for love. Is Tate perhaps our inquisitor of structures and our Gadamer? We may regard his inquisition of the sign.

As the somber wartime Christmas of 1942 approached, John Peale Bishop wrote to Tate and referred to their remarkable correspondence of the previous thirteen years. He remarked, "My Muse is silent," and

---

19. As noted above, p. 133, Lewis Simpson would have it so, notably in "The Critics Who Made Us: Allen Tate," *Sewanee Review,* XCIV (1986), 471–85; and "Eric Voegelin and the Story of the 'Clerks,'" *Sewanee Review,* XCVII (1989), 2–29.

20. Compare "Miss Emily and the Bibliographer," in *Reas.* (1941), 102–103: "Every point of view entails upon its proponents its own kind of insincerity."

continued in a wintry vein as if he suspected the fact of little more than fifteen months of life left to him: "There's nothing like old letters to recall the feel of times that are gone. I like from time to time to take yours out and peruse them, not only for the virtues they contain, but for the magic in them to give me myself again. I've felt in great danger lately of losing myself. The years seem to have stripped so much away. The wraith that's left seems thin and meagre" (*JPB*, December 23, 1942). Tate's letters were read to give Bishop not Tate but Bishop—letters with "virtues" of what Tate might well have called grammatical and logical or analytic kinds, shifting through wit, care, reflections of fellow feeling, and the like, to that "magic" of rhetoric which reinforced Bishop's presence to himself. Thus Bishop poses most of the issues with regard to Tate's sense and practice of the sign, poses them as the acute and sympathetic reflector who finds both more and less expectable mediation—more and less representational communion—in the signs, and who hints at a sign beyond the sign in the letters as epistles.

Ferdinand de Saussure's model of the sign will not represent the Tatian sign any more satisfactorily than Niels Bohr's early model of the atom—the dotlike nucleus with one or more circular orbits of "planetary" electrons—represents the dynamics of the atom. But there is a problem: those who have been virtually imprinted with the Saussurean emblem, so visually, sharply delineated as a two-story oval, with signifier above the floor-ceiling "bar" and signified below—a static emblem—may need alternative metaphorical models to defamiliarize the matter for fresh scrutiny and unprejudiced reflection.

The Tatian sign is a gerund, more like a cable carrying electric current or a pipe carrying a beam of subatomic particles than a two-dimensional oval with a bar, and more like a Möbius strip open to history and culture, an endless hermeneutic but not a solipsistic track. Culture and, with it, individual human identities have their being necessarily in mediation. One has one's self, like Bishop, by the dynamic intermediation of signs that are always another's. One becomes more oneself by talking to oneself in those signs and, one may say, those voices. Indeed, to avoid simplifying Tate's art more reductively than criticism is wont to do, one *must* say those voices. It is in such terms that it is easiest to make sense of his careerlong congeries of remarks about signs permuting and about signs beyond signs—or better, second-order signs—emerging. As early as 1931, in his review of Pound's *Draft of XXX Can-*

tos for the *Nation,* Tate identified the cantos as a model of community, an emblematic instance of, in Pound's words, "conversation between intelligent men," and prescribed "some years of hard textual study," which, he insisted with enormous significance, "is no more than our friends are constantly demanding of us; we hear them talk . . . and we return to hear them talk again . . . return for that mysterious quality of charm that has no rational meaning that we can define. It is only after a long time that the order, the direction, the rhythm of the talker's mind, the logic of his character as distinguished from anything logical he may say—it is a long time before this begins to take on form for us. So it is with Mr. Pound's cantos" (*Rev.,* 126). *Form* at the end of that remark pretty clearly means first- or second-order signs that can be handled and employed as mediums and be mediative for the self and others— including other versions of the self.

But they have to be handled tenderly, or as he would come to say by 1951, in "The Man of Letters in the Modern World," with brotherly love. By then, too, he would define the sign in communion, asked, sought, and cherished, by contrast with the sign in communication, the stamped mark or *typos* of power powering—hence his remark in "A Miscellany: Modern Poets and Convention" (1936) that "a new order of experience—the constant aim of serious poetry—exists in a new order of language." Any such order, almost an order of battle, or again *Ordnung,* is language formed to be mediative. In the famous phrase in "Ode to the Confederate Dead" about "language carried to the heart," he is drawing a contrast to "the moral conflict upon which the drama of the poem develops: the cut-off-ness of the modern 'intellectual man' from the world." That last phrase is from his essay "Narcissus as Narcissus" (1938), where he reaches out to gloss his public, but somewhat difficult, poem as an attempt to treat in a communal public way the narcissistic withdrawal from communion that has become pained and from power that has become defeat, an attempt depending in part on second-order signification so that the public form of the ode ironically becomes quasi-private meditation. If successful, "what was previously a merely felt quality of life has been raised to the level of experience—it has become specific, local, dramatic, 'formal'—that is to say, *in*-formed" (*Reas.,* 139, 144). Here the "level of experience" seems to be the intermediacy of signifying; felt quality is *experience* when one can signify it to another, although that signifying may entail perning in more than one hermeneutic loop, and will by the carrier (not, in false radio analogy, the receiver) be marginally permuted. The "versatility of translation is without limit" (1970; *Mem.*, 207).

Form and experience, whether of a new or a renewed or retranslated order, are to be achieved as literary signification by persons *ipso facto* always already implicated in history and culture, and necessarily as potentially, albeit approximately, interpersonal communion. Thus Tate with continuity from the thirties, yet with increasing clarity after 1940, reinforces and glosses with respect to the dynamics of the sign the decreasingly positivistic and increasingly inclusive ideological position Mumford somewhat inconstantly promulgated. He also illuminates the rationale of Eiseley, who was less pointedly implicative on the sign, though plain enough in precept and practice on language and communion (see below, pp. 189–99).

It is worth remarking that Tate wrote as if in substantial accord with Jan Mukarovsky, whose writings in Czech began to be reported in English by René Wellek only in 1946 and 1949.[21] In the journal of the Prague Linguistic Circle, Mukarovsky's long essay "On Poetic Language" (1940) insists that "the linguistic sign is actually symmetrical with respect to reality. The sound aspect proceeds from reality, the semantic aspect tends toward it, though only through the mediation of psychic phenomena (images, emotions, volitions evoked by speech)."[22] Whether consistently with Mukarovsky's intentions or not need not be of much concern here, but his "from . . . toward . . . mediation . . . evoked by speech" can stand as an intermediate summary—a kind of subtotal—of Tate's emerging emphasis on the sign as referential and based on reality both internal (to the self) and external (of the nonself), in a dynamic interpersonal loop subject to change.

Just as significant and apposite, Mukarovsky went on in the same essay to emphasize "semantic statics and dynamics" as together creating the "basic antinomy of every semantic process," and to emphasize the literary work as a formal yet concrete "semantic gesture," cultural in

21. William S. Knickerbocker, *Twentieth Century English* (New York, 1946); René Wellek and Austin Warren, *Theory of Literature* (New York, 1949). Tate surely saw the latter book.

22. Jan Mukarovsky, "On Poetic Language," in *The Word and Verbal Art: Selected Essays of Jan Mukarovsky,* trans. and ed. John Burbank and Peter Steiner (New Haven, 1977), 18.

ways including an author's personality, without however yielding to deterministic psychology. And he granted that "dialogic and monologic qualities are in fact both present at the very origin of every utterance" (pp. 47, 54, 64, 58). Just so might Tate have acknowledged the *effectual* presence to the self, mediated by the more or less compound sign, of an other (Nastasya, *semblable, frère*)—never utterly present but not necessarily lost to *différance*.

To reiterate and reinforce, in an expanded loop: Tate practices and partly enunciates in the later essays that are foregrounded in this chapter a supra-Saussurean position, largely that which Colin Falck has recently delineated with elegant concision. The Saussureanism Falck attacks may serve as a foil here. Falck controverts both Saussure's Principle I, that the sign is by nature *arbitrary*, and his Principle II, that the sign signifies only relationally, with the relations being other signs and language a system of differences. His complaint—fairly made against widespread views, I think, even if not so fairly charged against Saussure or Derrida as Falck believes—is that the two principles have become a license for the unnecessary and indeed untenable notion that the relationship between language and world does not entail or even admit a "kind of 'naming process' in any sense whatever." He holds that many a structuralist and poststructuralist has found it impracticable to discuss the referentiality of signs to world in Saussurean terms and has even deemed it undesirable to discuss referentiality in any terms.[23]

That is true of some analysts, no doubt, and it defines Tate—not to mention Mukarovsky—by contrast. The living body in the republic of letters or by the sheeted bed is emphatically the biologically and culturally embodied self Falck wants accounted for yet finds occluded by usages of *langue* and *synchrony* that demote, not as a reasoned conclusion but by definitional fiat at the beginning, *parole* and *diachrony*. Falck champions what he calls extralinguistic reality, meaning roughly both prelinguistic bodily senses of being and what might be called aspirations to and intimations of transcendence.

So does Tate, but here a different contrast becomes important. As "The Hovering Fly" and "The Man of Letters" suggest, each with its

---

23. Colin Falck, "Saussurian Theory and the Abolition of Reality," *Monist*, LXIX (1986), 133–45; reproduced in *Myth, Truth, and Literature: Towards a True Post-Modernism* (Cambridge, Eng., 1989), Chapter 1.

different emphases, he honors what others have called the wisdom of the body and honors the urge toward transcendence. But he also honors the sign as a human semicompletion of what might be called protolinguistic and semilinguistic reality. The Tate who early on talked about religion and the "whole horse" and who seems to have been less interested than Mumford or Eiseley in the dawn of culture and language would perhaps have countenanced a baseball analogy to define his sense of the sign, and even the poetic sign (although televised football was his preference).

It is possible to imagine a fellow I'll call Abner Doubleday, who knew some outdoor games employing various spaces and equipment and exertions, hefting a stone and gradually coming to think, "Better for casting than a tennis ball . . . call it . . . call the player who throws it both for and against the batter . . . ," and so on. Moreoever, the pitcher's unmistakable pitch moves and has its being in semilinguistic incompletion until it is ruled, that is, competently named, a strike or a ball (occasionally even a wild pitch) or else is competently named to have hit the batter or to have been hit fair or foul or for an error. The man of letters, we have seen Tate assert, names competently the phenomena everyone can in some partial sense experience as going on. The prophet of letters, if the analogy will bear another turn, competently names by uttering the megaword, not truer than *strike* or *fair ball* but more definitive—say, *fixed* or *drugged;* say, *myth* or *positivism.* And, in a conflation of umpires with players, it is possible to imagine Doubleday and it is proper to imagine Tate—who may not have played baseball but did play the violin—as feeling, and in part linguistically thinking, the well-doing of their completions to be a delight and discipline to the self and soul.

One kind of ill-doing of such naming lies in what Tate complains of, in "The Man of Letters" (1952), as the "staggering abuses of language, and thus of choices and ends, that vitiate the culture of western nations" (*Col.,* 391). There is also meta-ill-doing, as in what Falck calls the Saussurean reduction of language to a comfortable object of study equivalent to logic. Tate complained as early as 1936, in "The Function of a Critical Quarterly," of "our own splitting off of information from *understanding,* this modern divorce of *action* from intelligence . . . *trying* to think *into* the *moving* world a rational order of value" (*Col.,* 63). The emphasis is mine, but the classically plentiful gerunds and participles are Tate's emphasis on dynamism, historicity, and implication—against, on the one hand, the tychastic world of vicissitude and, on the other, the mechanistic world of taxonomy.[24]

Tate later stigmatizes apocalyptically the grammarizing and logicizing of the world as leading to the last carp in the pond (see above, p. 179). Indicative of his increasingly resonant naming is his indictment in 1952 of a variant abuse of the contingent and permeable and implicated sign: "that idolatrous dissolution of language from the grammar of a possible world, which results from the belief that language itself can be reality, or by incantation can create reality" ("The Angelic Imagination: Poe and the Power of Words" [subtitle altered in 1953 to "Poe as God"], *Col.,* 437). His focus in that essay is Poe and a lineage including Mallarmé, his own friend Hart Crane, and Dylan Thomas. But the indictment scarcely exempts any who would divinize themselves or at least sacralize their discipline by purporting to make language an object of study like logic. The temptation can be acknowledged: if language is a machine, the mechanic is a god.

For Tate, the evil of mechanistic language—applied Saussureanism, in Falck's sense—is not that language is conceived as relational (in the way the parts of a machine are by definition relational) but that it is not conceived as relational enough. We need, he both argues and demonstrates, to construe not *the* sign nor even signs among or differing from others in a quasi-static network of other quasi-unitary signs but, rather, and more classically—in the terms of his Longinus—signs relating in a field or, even more intrinsically than that, in a *mode,* of ethos. And ethos is intrinsic to human selves as they inevitably feel, evaluate, and make themselves present.[25]

Ethos cannot be eliminated from the sign that is necessarily in action necessarily among signs, although it can be occluded by the reportorial

24. See also, in "Literature as Knowledge," Tate's suggestion that a positivistic proposition is dangerous in excluding knower and knowing as radically as do—formally, at least—the terms of a chemical equation (*Reas.,* 27).

25. Throughout "The Unliteral Imagination; or, I, Too, Dislike It" (*Southern Review,* n.s., I [1965], 530–42), Tate's complaint is of the "unliteral" as the insufficiently *relational.* In one example, J. A. Carlyle is faulted for mistranslating Dante's *l'ombre* in *Inferno* xxxiv not as *shades* but as *souls.* Tate's point is less directly a matter of theology than of the nature of the sign: *souls* obscures figural Vergility and the presencing self, that is, the mind apprehending "*essentia* as distinguished from *existentia.*"

mode, or cruded down to an ethos of power or the will to power, as it is in communication in the sense figuring in "The Man of Letters": type stamping, manipulation. In "Longinus and the 'New Criticism'" (1948), he notes that "our multiverse has increasingly, since the seventeenth century, consisted of unstable objects dissolving into energy . . ." (*Col.*, 527). Whether by argument, prejudicial definition, or mere assumption, to detach ethos from the sign as if ethos were extrinsic to the sign and were mere *parole* or any such quasi-reductive construal as diachrony, is for Tate as unwarranted and humanly dysfunctional as the eighteenth-century demotion of allegedly secondary qualities. The neglect of that truly significant mode of the system of difference animating linguistic interchange—*ethos* of force/*ethos* of love—may be the preeminent special case of the eighteenth-century and modern cultural catastrophe.[26]

In "Longinus' profound but topical dialectic" of subject matter and occasion, occasion is "the relation between the poet and the person to whom the poem is addressed . . . a sense of the 'point of view' . . . the *prime literary strategy* which can never be made prescriptive" (*Col.*, 515, 512; my emphasis). When Longinus has concluded his discourse on occasion, technique, and nature, saying, "Most important of all, we must learn from art the fact that some elements of style depend upon nature alone," Tate goes on:

> Most important of all, I make Longinus say, we learn from the development of technique that stylistic autonomy is a delusion, because style comes into existence only as it discovers the subject; and conversely the subject exists only after it is formed by the style. No literary work is perfect, no subject perfectly formed. Style reveals that which is not style in the process of forming it. Style does not create the subject, it discovers it. The fusion of art and nature, of technique and subject, can never exceed the approximate; the margin of imperfection, of the unformed, is always there—nature intractable to art, art unequal to nature. The converse of Longinus' aphorism will further elucidate it: we must learn from nature that some elements of subject matter, in a literary work, "depend" upon art alone. There is a reciprocal relation,

26. As early as 1934, in "Three Types of Poetry": The "modern scientist" is the "more efficient allegorist of nature," and in allegory "the image is not a complete, qualitative whole; it is an abstraction calculated to force the situation upon which it is imposed toward a single direction" (*Reac.*, 92). This is no doubt unfair to some allegory and unfair to some scientists, but *force* and *impose* start to make the critical distinction about the religion of techno-power.

not an identity—not, certainly, the identity of form and content—a dynamic, shifting relation between technique and subject; and they reveal each other. This is my sense of Longinus' primary insight. It is an insight of considerable subtlety that has a special claim to the attention of our generation. (*Col.*, 516–17)

So the ethos-infused style or compound sign—and the sign is essentially compound—strives in a reciprocal relationship. Like Adam, it makes that which was not humanly in nature until named—and *nature* is one of the names, as Eiseley would argue a dozen years later. It wrestles something of the extralinguistic into community, which is essentially though not exclusively linguistic—something of the semilinguistic, the prelinguistic, even the transcendent.

For Tate, as for Dante and the Renaissance poets he admired, there is an unlikelihood that we will complete the creation of matters since Eden and Babel by naming them *quite* according to their nature. To be in the world humanly, beyond the most limited and amorphous feeling, is to be experiencing in mediatively linguistic fashion in conjunction with others. And *parole* and *langue* will be a hermeneutic journey, at best, subtotals. The great poet of prose or verse will signify matters old and new more nearly according to their currently experienced nature, and thus more decisively—*parole* amending *langue* as used, diachronic *langue,* not *langue* dead in abstraction.[27] For abstraction is, we have been told, the death of religion and of everything else. The Saussurean notion of *langue* is abstractive, leaving language essentially a meaning machine and only accidentally a joint venture of persons speaking in history. But the republic of letters is no Hegelian state; rather, for Tate, it is a company of literate neighbors with more or less effective representatives and countless domiciliary householders, simultaneously lodged and on the move.

27. For a familiar deconstruction of the quasi clarity, the pseudodecisiveness of typography, of signifier and of signified, emphatically in the Derridian ethos, see Jacques Derrida, "Living On. Border Lines," in *Deconstruction and Criticism*, ed. Harold Bloom *et al.* (New York, 1979), 75–176. For closure on the fundamental matter of orality with which the consideration of Tate began, and Samuel Johnson's degree of awareness that "common sense" and the "common reader" were not concepts as tidy as print, see Walter Ong, "Samuel Johnson and the Printed Word," *Review*, X (1988), 97–112.

6

ALL THAT IS CRUSHING US

*There is nothing so illicit as to dwarf the world by means of
our manias and blindness, to minimize reality, to suppress
fragments of what exists.*

—José Ortega y Gasset, *Meditations on Quixote*

Mumford may be described broadly as a student of technological urban civilization who started generally enchanted though with particular complaints and became the prophetic opponent of the myth of technology, though with a surviving appreciation for technological services to life. With Tate, there was a less concretely particularized, a more intellectually thematized, consideration of the rural and the urban, the less literate and the more literate, American social fabrics. His progress to his prophetic stance was in an obvious sense more spiritualized than Mumford's. Loren Eiseley approximately recapitulated Tate's development. But it is convenient to join Eiseley farther down the road to prophecy, later in the development of his own powerful ethos, when he arrived out of digs, illnesses, and personal anguishes to view the American cultural project and its history, indeed *Homo sapiens,* as *non sapiens* and instead a not yet determinate episode in geologic and biologic time. This chapter perforce arrives in the middle of things and, appropriately, I think, makes a small hermeneutic loop, leaving expansions to the two following chapters. But first, here, metaxic Eiseley, in between knowing and not knowing, telling and not telling, life and death; then, cultural idols; and finally, problematics of prophetic relationship.[1]

1. I employ *metaxy* as it is renewed from Plato's *Philebus* and *Symposium* by Eric Voegelin (*Anamnesis* [Notre Dame, Ind., 1978], 103). My title phrase is in Eiseley's interpretation of the "talking cat," in *All,* 228.

Eiseley took as his epigraph for *The Unexpected Universe* the remark of J. B. S. Haldane that "the world is not only queerer than we suppose, but queerer than we can suppose." Eiseley agreed, too, though not epigraphically, with the Pauline and Augustinian notion that the iniquity in the heart of man is a mystery, not a mere puzzle to be teased out into causes, components, and effects or remedies. His life in writing, as in field and classroom and library, may be conceived as a labor of ambivalence: to get the nameable called by the right names and the unnameable called by serviceable names, and to show and tell alternatives to manic dwarfings, skewings, and suppressions, whether those would obscure the benign or the malign. A "toddler . . . pleading wordlessly for . . . peace" between quarreling parents, he knew the "feeling they represent," which is expressed in such adult words as "terror, anxiety, ostracism, shame" (*All*, 24). Any effort to recognize Eiseley's practice of prophecy will involve one, as his efforts involved him, with words lexically and syntactically and ideologically implicated in all he found repellent in the twentieth-century American version of the Western world.

Haldane's notion of the world as "queerer than we can suppose" has at least two corollaries vital to Eiseley. The first is that the languages different specialists, or indeed nonspecialists, use to talk about the world may occlude one another or may fail to connect, leaving crucial gaps between languages. The second is that any language may leave gaps between itself and human experience of other than linguistic and not yet linguistic forms. And that is not even to mention reality as God might be supposed to know it, a separable perplexity for Eiseley. Even the phrase *about the world* prejudges the issue, as he realized, by suggesting a kind of circumscribing more neatly definitive than discourse can validly pretend to be: "Language . . . creates for man an invisible prison. Language implies boundaries . . . a rabbit, a man . . . frozen into a concept, a word. Powerful though the spell of human language has proven itself to be, it has laid boundaries upon the cosmos" (*Inv.*, 31). This may be taken as Eiseley's most explicit comment, generally true to his attitude after *The Immense Journey*, about the provisional and potentially reductive relation of signifiers to signifieds and to referents. The convention of indication, of adequacy, between signifiers, signified, and referents, may be a seriously misleading fiction he insists in such paragraphs and in his essay "How Natural is 'Natural'?" in *The Firmament of Time*. So may be the convention of arbitrariness, mere difference, nonindication.

He knew that every element of signification—signifier, signified, referent—can derange the others and be deranged by the flux of history. In a passage in which almost every word might bear the ironic quotation marks he saved for *expected*, he reflected on the cosmos as a theater and as such (in a term he acknowledges from Jean Cocteau) a "trick factory": "Man is no more natural than the world. In reality he is . . . the creator of a phantom universe, the universe we call culture—a formidable realm of cloud shapes, ideas, potentialities, gods, and cities, which with man's death will collapse into dust and vanish back into 'expected' nature" (*Inv.*, 120). He obviously questions *natural* and *culture* as signifiers, as signifieds, as enduring referents, and thereby *man* as well. In the essay after the one in which he characterizes the cosmos as theater he cites José Ortega y Gasset to the effect that "there is no human nature, there is only history" (*Inv.*, 144). History, like signification, is a function of dynamic human interrelationship, always risky and uncanny.

The moments these remarks of his give voice to are at or close to the extreme of his explicit particular concern with the medium of signification. His more usual concern was with a larger dynamic of interaction: languages or vocabularies or realms of discourse, options for action, human feelings interacting. He warns of "dangers implicit in any great urban civilization whose hold on the reality of its association with nature is slipping away." He means to include what both is and is not his own society, and he explains: "There is . . . a difference between the mastery of nature and respect for the powers of nature that are unmastered by man. The first attitude leads to exploitation and ultimately to destruction of resources, the latter to a protective ethic of respect that primitive man has frequently understood better than our technologically oriented society." [2] Here the focus may be on emotionally loaded signifieds wrenching or occluding signifiers and referents alike, with enormous practical consequences.

That a hold on reality or an attitude is linguistic—perhaps, for Eiseley, essentially linguistic—becomes clear in moments like his characteristic expression of envy for the "dark-faced old shaman" who reads the cracks made by a campfire in the shoulder bone of a rabbit: "He knows about the daily renewing of a pact with man; we hear nothing / but the echoes from a deserted universe" ("Pioneer 10," in *Inno.*,

2. Loren Eiseley, Introduction to *The Shape of Likelihood: Relevance and University*, ed. Taylor Littleton (University, Ala., 1971), 5–6.

22).[3] Thus, to speak of discourse about the world or of gaps between languages or between some language and some world to which it might be supposed to refer is to touch on matters deeply ambivalent for Eiseley. Writings may benignly fix meaning, and there may be a daily renewed "pact with man" in the shaman's rabbit bone as in more familiarly scientific discourse, as contrasted with the evanescence of whispers, echoes, vocal words on the wind. But the fixity of representation and spatiality of conception is gained at the expense of occluding most of the dynamism in experience and may be only an apparent fixity. The natural world treats spatial structures roughly: coyotes, in a late poem, "are the echoers, we / a jumble of leaves and dust / quickly gone by. Lovers of form we will be formless / in the tales to come" (*Ano.*, 38). These "tales" he imagines to be quite inaccessibly alien, which implies that the echoes of them by so long enduring a species as the coyote are doubly alien. Rather similarly, he elsewhere imagines—shamanizes— readings of signs by such nonhuman beings as a spider, a fox cub, a stag, corpuscles in his own blood, cacti. But reciprocally to the way such persistent orders of being lose in sophistication what they gain in species endurance (and lose, *a fortiori*, in individual endurance), hot human culture loses in mutability what it gains in sophistication of messages:

> I know now what impulse created the Olmec heads
>   Mayan stelae and Machu Picchu.
> The stone will survive us. Like the Old Ones I
>   have left a shaped stone in the gravel.
> It is all my knowledge. It will lie there when no
>   one interprets these words.

> ("The Little Treasures," in *Ano.*, 57)

"*All* my knowledge"—compression and reduction to essence, not unlike Yeats's saying in "Of the Coming of Wisdom with Time," "Now I may wither into the truth." But Yeats (to pursue the definition into contrast) proposes words "to be written on a stone at Thoor Ballylee," and petitions, "May these characters remain / When all is ruin once again." He knows there is no guarantee that they will not, too, be despoiled. But Eiseley acknowledges a still later time when anything whatever written in the characters of human alphabets will be meaningless in any human sense, as *human* is at present understood, and he seems to discount his

---

3. Eiseley changed the title to "Pioneer 11" for subsequent editions.

knowledge, all of which can be reduced not to a microchip but to a stone.

So there is an ideological issue about the language he must use to be understood by himself and us, and an issue of mutability with regard to that language's accessibility to a future audience. Both these points must be returned to. There is also a structuralist concern, which may be put most clearly and comprehensively not by any quotation from Eiseley's published writings or unpublished writings that I have seen, nor from the texts of Claude Lévi-Strauss, which he owned, nor from Edward Said's early book on Joseph Conrad, which he owned, but rather by a bit from Said's brilliant chapter "Abecedarium Culturae" in *Beginnings:*

> The structuralist substitutes order, or the structure of thought, for a Being that in classical philosophy had informed and nurtured thought. Order is a limit beyond which it is impossible to go, and without which, moreover, it is impossible to think. Order is the result of the mind choosing syntax over semantics. . . . The structuralists, in short, do not believe in the immediacy of anything: they are content to understand and to contemplate the alphabetical order of sense as a mediating function rather than as a direct meaning. . . . To perceive this order one cannot have recourse to a direct unfolding (as in the *Entfaltung* of hermeneutical interpretation). . . . That alternative . . . disappeared with the primordial Origin. We are left only with . . . experience conceived of as a gigantic script or musical score . . . [a] search for structure as *Zusammenhänge,* the "principle of solidarity" among parts, according to Barthes.[4]

Eiseley's style in language and life, and his fascination and intractability, inhere partly in his insistence on both syntax and semantics. The discipline of paleontology, which he embraced, employed, taught, and enhanced, may be thought of as structuralist in Said's sense and in Jean Piaget's triadic formulation of structuralism as totalizing, self-referential, and transformational. Eiseley both acclaimed *and resisted* paleontology's spatializing and taxonomizing of geological and biological process, its totalizing. He was skeptical of the faith that the order of human sign systems is a limit beyond which it is impossible to go or think. The syntax of paleontology, after all, mediates the dynamism of the thinking process with conceptual entities globalized from areas of early identification. Thus the Silurian level lies not just relative to other strata in Wales but identically relative to those strata under North Ameri-

4. Edward Said, *Beginnings: Intention and Method* (Baltimore, 1975), 326–27.

can feet, and does so conceptually even if local geological disturbances have erased or scrambled it, or worn it away, leaving us walking anomalously on a lower level, with the Silurian formally in midair above us. And under the Silurian level is the Cambrian, everywhere, not just under Cambria. Far above it will be identified, in any longitude at sufficiently northern latitudes, evidence of Würm or Günz or Wisconsin glaciations, and at whatever latitude, indications of the Pennsylvanian period—a name that in geological contexts startlingly disintegrates and fossilizes the bits of linguistic and sociopolitical history inhering in the initial meaning "pertaining to Penn's forest."

But Eiseley's metaphors do insist repeatedly and strikingly on the dynamic unfolding of being. And a recurring word like *far* in his usage will tend to mark an inadequacy or breakdown of totalizing attempts by any system of language and mark such breakdown or failure even at the risk of sounding vague or mawkish. Indeed, his poems, essays, chapters, and books all have some kinship with that shaped stone to be left in the gravel—"all my knowledge. It will lie there when no one interprets these words." Part of his knowledge is the centrality of the human impulse both to form or differentiate and to reach out affectionately to another in communion, even in indecipherable communication with an almost unimaginable other. Walter Ong has clarified that, as a title of his puts it, "the writer's audience is always a fiction." [5] Eiseley sometimes seems to be imagining two audiences: one conceived to be siblinglike, akin to himself and ourselves, the other so unlike ourselves as to be difficult to conceive.

For him, the word or stony sign may variously mediate presence or be presence, not presence of the godly origin, necessarily, but presence of the "impulse that created the Olmec heads" or other impulses such as the grief of Stone Age parents at the grave of their child or the aspiration of Eiseley himself to define a perennially emerging self. Is the shaped stone mediative or immediate for the sympathetic finder? One understands that the stone can be purely instrumental to the finder sufficiently intent on finding a handy missile, or a convenient shape for a mosaic, or a makeweight. But is that all?

Does one say the words are adequate, or mediative, or immediate, when one of Eiseley's numerous early elegiac love poems says, "No

5. Walter Ong, "The Writer's Audience Is Always a Fiction," in *Interfaces of the Word* (Ithaca, N.Y., 1977), 53–81.

peace till it's gone, / Till you think of her warm mouth / With a mouth of stone" (*Wings,* 39)? It may be as difficult to conceive that pain apart from its verbal representation and prescription as it would be to conceive the game of golf apart from clubs, strokes, course, and balls. But Eiseley himself can stigmatize bookish descriptions, asserting their inadequacy, as in his curious poetic mélange "Habits Nocturnal" (*Ano.,* 36–37), where he presents the being of nocturnal and endangered species as better understood in cerebral and visceral senses by the nocturnal and endangered Eiseley than by zoology. As will become plain in detail, a crucial part of his prophetic message is explicitly and implicitly, by mediative argument and immediate witness, to deny the efficacy of any system that would presume to freeze-dry or totalize.

On the other hand, paleontology is an empire intrinsically without guilt, though admittedly a technician hustler like the one in "Obituary of a Small Bone Hunter" (*Night*) can import culpable action to it. Even at that, the history of human aggressions is dwarfed, thus far, by the great planetary extinctions. We win paleontological knowledge from nature, as in a larger way Darwin did, with little of the moral and fleshly risk Conrad epitomizes in Kurtz. But one recalls, and may suppose Eiseley recalled, that there was a different sort of risk to which Conrad's Marlow was vulnerable. There he sat, the mediatorial tale-teller on the deck of the yacht—like Eiseley or the prophet Amos—no longer at ease in Zion or Thames estuary or campus office.

The structuralist would tend to believe that, as Said puts it, "the categories of thought and language are identical."[6] I do not know if Eiseley was aware of Alfred North Whitehead's report that he sometimes thought in concepts, or how telling a counterargument he would have taken that to be, given the mathematical and quasi-mathematical direction of much of Whitehead's thought. But on the published record, he attended with profound respect to the partly nonverbal imagery of dreams. Although his late essay "Science and the Sense of the Holy" (in *Star*), breaks with Freud on matters of the "oceanic" feeling, he seems to have honored Freud as the prophet of the unconscious, even speculating in one note to himself (in the archives at the University of Pennsylvania) that Freud might serve as an organizing focus for his future books. And one reads his verbal reports on seemingly spontaneous and nonverbal actions, such as gestures of love toward all manner of crea-

6. Said, *Beginnings,* 316.

ture, as well as quasi-verbal gestures like making and leaving (slightly) formed rocks in the gravel. One observes that he owned both *Personal Knowledge* and *The Tacit Dimension* by the late Michael Polanyi. Eiseley so esteemed books as expressive artifacts that he almost never marked them; unless a dust jacket or cover is worn or he acknowledged a book in the somewhat sketchy bibliographies to his own books, it can be unclear how intently he read even a text we know he owned. But there are few modern authors for whom he owned more than one work; and Polanyi's argument for a wide range of knowledge, representation, and self-presence beyond the verbal range was at the very least a congenial position for Eiseley.

"What, then, is to be done?" Language in both general and special realms of discourse tends to misrepresent or to fail of connection, and privileged moments of immediacy may be as problematically shareable as Wordsworthian spots of time.[7] The first answer, at least roughly primary in Eiseley's becoming the writer we partly know, seems to have been an act of quite teacherly statistical faith. Those who teach may think that when the classroom is peopled with university students, they will exhibit for historical even if not for ontological reasons a range of awareness and interest and experience substantially akin to the teachers' own. It has to be a guarded act of faith: one knows that some students will only be there—and there only minimally—for administrative or parental reasons so deeply resented that they will be inaccessible; some will be disabled by individual or familial histories or circumstances more melodramatic than Eiseley's; some, perhaps most, will not have become responsibly (verbally?) *conscious* of the interest, feeling, and knowledge they do have. Some may even be living in one or another world so private as to qualify as insane. But as a practical matter one worries less about homicidal maniacs than about lacking the nimble control of one's subject matter to dispose and articulate it for the connections presumed to be present in that audience. It was exactly so when, at the beginning of the formal teaching component of Eiseley's career, he was assigned to teach introductory sociology. Anthropology or paleontology he would presumably have taught less anxiously, with what could in a sense be called a more complacent reliance on mere subject matter. His account of the experience implies that he tried every-

7. After writing this, I was pleased to find Andrew Angyal construing the essays as Wordsworthian spots. See his *Loren Eiseley* (Boston, 1983), 93–94, 99.

thing: "Like the proverbial Russian fleeing in a sleigh across the steppes before a wolf pack . . . everything from anecdotes of fossil hunting to observations upon Victorian Darwinism were being hurled headlong from the rear of the sleigh" (*All*, 131).

One reaction to an essay or chapter by Eiseley is likely to be the feeling that he is trying everything with the reader. When chronological sequence does not govern the structure, and sometimes when it might, there can be difficulty in descrying a structure, much more in outlining it. There is deliberate variety; there is nearly always at some point an *event* that was resonantly meaningful for him and therefore (on the basic faith assumption) that he trusts will be so for most of the people he addresses. A scaffolding of connections is thrown up, seemingly on the rationale that his readers may not have brought their own but will listen for a time to one who might say with Whitman, "I am the man, I suffered, I was there."

Another declaimer may be invoked for comparison. An acute critic of Milton has remarked of him, "Master of the conventions of public discourse, this man excelled in the rare art of speaking intimately before an audience."[8] That is Eiseley, right enough, but with a difference, because our age is even more disjunctive than Milton's. Said has understood that "when words lose the power to represent their interconnections—that is, the power to refer not only to objects but also to the system connecting objects to one another in a universal taxonomy of existence—then we enter the modern period."[9] So Eiseley struggled vocally and textually to make connections among the fractionated mental landscapes, the masked and occluded horizons, the disjunctive realms of discourse at the University of Kansas, Oberlin College, the University of Pennsylvania—wherever around late-twentieth-century America he might find that "fit audience" Milton finally sought.

In many a college class he probably began literally with an appeal to a defined and presumptively shareable sensory complex. In most essays and poems, he begins figuratively or implicitly with such an appeal. His substantial beginnings are often marked with a demonstrative pronoun or adjective: *this* object, *here*, in *this* place, *this* scene or trail of particulars, *that* topic or question. The rationale is manifold for this show-and-tell maneuver: our sensoriums may be expected to vary less than our

8. William Kerrigan, *The Prophetic Milton* (Charlottesville, Va., 1974), 11.
9. Said, *Beginnings*, 285.

histories, the sensory gestalt he represents in words may serve in a disjunctive world as a local taxonomy of existence as if it were a camp made or tent pitched, and the degree of *control* is high.

Chapter 8 will focus more expansively on the movements of discrete writings by Eiseley. But at present, in the context of introductory observations on the linguistic medium and its cultural circumstances, there is need for a moment's deliberation on control. Control was urgent, for Eiseley. He feared "chaos"—not a word he would use casually—in his introductory sociology class. College students can be a pertinacious lot—but in Lawrence, Kansas, in the late thirties? He usually stood to teach, a posture that tends to put the teacher at a greater physical and existential distance than being seated does, while making it easier to fill the space of the room with a dominant combination of voice and moving presence.

Voice is a "summons to belief," as Walter Ong has argued, and we must "style writing a 'secondary modeling system,' dependent on a prior primary system, spoken language."[10] Yet Ong goes on to argue, for intensely literate persons in general, what seems particularly true of Eisely: "To dissociate words from writing is psychologically threatening, for literates' sense of control over language is closely tied to the visual transformations of language."[11] There is an unusually close parallel between the writer-up of occasional moments of intimacy, about which he was ambivalent, and the lecturer choosing such data to cast from the back of his fleeing sleigh. That the analogy has students as wolves sufficiently attests to his ambivalence—good-naturedly jokey but even so real—about the enterprise. When one scholar broke away from Eiseley's own design for a course on Eiseley (apparently conceived as "lectures, prescribed readings, limited colloquia") in the 1970s, the subject of the course evidently became quite angry. The point that seems to have disturbed him most was the possibility of uncontrolled talk with former associates.[12]

10. Walter Ong, "Voice as Summons to Belief," in *Literature and Belief,* ed. M. H. Abrams (New York, 1957); Ong, *Orality and Literacy: The Technologizing of the Word* (New York, 1982), 8.

11. Ong, *Orality and Literacy,* 14.

12. See E. Fred Carlisle's indispensable *Loren Eiseley: The Development of a Writer* (Urbana, Ill., 1983), Preface and Chapter 11. With reference to *control,* see, too, Carlisle's remarks on science as a control welcome to Eiseley for both his inner and his public life (pp. 69, 112).

He was markedly less eager to control nature than he was to avoid control by the "fairy ring or magic circle" of one or another human system (*All*, 187). It is, of course, a commonplace of systems analysis that one man's system is another man's subsystem, and Eiseley's deepening concern over the last forty years of his life was—or at least his orientation was toward—an inclusiveness so profound that hitherto-formulated human boundary projections fade. His reasons were the Ortegan ones that stand as epigraph to this chapter. He would, I believe, as an approving citer of Kierkegaard on the future, have endorsed the further point that, as William Spanos, Paul Bové, and Daniel O'Hara put it, "We also know from Kierkegaard that the desire *to make, to will* new beginnings rather than *letting them occur* in the context of their occasion is a repeated self-enticement for the Western consciousness." [13] It is an uneasy, even precarious, standpoint one occupies if one would advance scientific thought and, like Eiseley later, what may for the moment be called cultural thought but is anxious lest one force beginnings, or if one collects occasions from fieldwork and reading, collects them in mind and notebooks, but is manifestly anxious lest they be forced by others while still in embryo or lest they be interrogated intrusively in their public appearance.

Profound kinships between the Eiseley of platform and printed page do appear, to certain modes of reflection. But for no one more than for him was "the writer's audience . . . always a fiction" and his writing a removed and solitary business. Seemingly quite early in his scientific career, he emerged from a death-defying descent into a cave with a permanently altered and alienated perspective (*All*, Chapter 10). The first chapter of *The Immense Journey*, his first book, begins face-to-face with a skull so deep in a geological crevice that there is scarcely a slit of light from the sky visible. Here are neo-Homeric journeys to the underworld. His scene of writing, he suggests (whatever his biographical statistics would have shown), was normally the cave of night, conventionally the place of "death's second self," sleep. But for the sleepless Eiseley the hours of dark were that quasi tomb of consciousness, the "night of time" (*Inv.*, 91). He who in his highly selective but multiple autobiographical writings recounted almost actualized deaths or symbolic deaths even prior to the one in the cave would seem to have come, al-

---

13. William Spanos, Paul Bové, Daniel O'Hara, eds., *The Question of Textuality* (Bloomington, Ind., 1982), 4.

most like Lazarus, to speak as if from the tomb or from a night journey, with a knowing and fascinated dread of death and a disdain for usual human boundaries and enclosures.[14] He makes much of writing not only at night but in hotel rooms, where "tomorrow, the shadow on the wall will be that of another" (*Night,* Preface). One reference highlights the cryptlike quality of his study, the "book-lined room in which I write" (*Inv.,* 93). Another, almost the only in his mature work that explicitly makes writing a daytime activity, seizes on the great solar eclipse of March 7, 1970: "Beyond my window I can see [from my desk] a strangely darkened sky, as though the light of the sun were going out forever. For an instant, lost in the dim gray light, I experience an equally gray clarity of vision" (*Inv.,* 69). Did he actually draft the passage at that desk in those brief moments of the eclipse? Not improbable. Did he later go over it with at least an eye to revision? All but certainly so, and available manuscripts of other writings suggest he probably altered occasional words and phrases. No writer would suppose otherwise. What we know with logical and formal certainty is that we are reading after the fact, after the complex fact of a twilight that he defines in the context of the excerpt as an emblematically sepulchral twilight, but that he defines so for the inhabited world.

For my analytic purposes of clarifying Eiseley's rationale of capturing the flawed and provisional liveliness of teaching talk in print, I have reversed his own discursive order, which is more polemical and aims to clarify the fragility of the cultural world: "[Since the world's] resources are not to be tapped without the drastic reordering of man's mental world, his final feat has as its first preliminary the invention of a way to pass knowledge through the doorway of the tomb—namely, the achievement of the written word. . . . The dead can be made to speak from those great cemeteries of thought known as libraries" (*Inv.,* 63).

There is cause to ponder, and return to, the microapocalypse of someone separating self from self and any other by writing, always, but potentially by the same act transmitting life. Eiseley asserts the transmission, and then moots it with a version of great apocalypse.

---

14. I've been moved to some steps taken in this paragraph by Fred G. See's "'Writing So as Not to Die': Edgar Rice Burroughs and the West Beyond the West," *Melus,* II (1984), 59–72.

Eiseley strenuously achieved an exemplary position with regard to his profession and indeed to contemporary American society. He willed that position to be a model; he presented the achieving of it as a kind of argument. The qualities and components of emerging American society, and the steps to his mature position, become subordinate terms of his lifelong argument. He did not quite nail ninety-five theses on the door of the National Academy of Sciences, yet in at least a double way his writings argue for a reformation and against the scientistic, technologistic culture he saw as increasingly characteristic of what is often and vaguely called the modern world. Of course he argued in concert with others he had read, such as George Santayana and Ortega y Gassett, and his admired colleague Mumford. Indeed, from the late 1940s onward he may be said to have been dilating on a conviction of Pascal's which he quoted in 1960 (*Fir.*, 159): "There is nothing which we cannot make natural. . . . There is nothing natural which we do not destroy."

All the Strange Hours: The Excavation of a Life, although not the final book published under Eiseley's name, was his most forthrightly autobiographical.[15] Its title suggests the necrological, almost the necropolitan, and suggests the historical past or even pluperfect tense. All of that masks the drama of a book that begins with an affront in the form of a second text, typographically highlighted:

> Anybody who objects to the sight of Nude People Making Love doesn't belong in here. Anything San Francisco can do we can do better. (*All,* 4)

That sign on the door of a Texas bar tests the reader to choose: to join Eiseley in his reaction to some degree and to continue in his text with him, or to be indifferent to the sign's offensiveness to him and by that much an outsider to his text even if an ongoing reader of it. The test lays out modes of relating, not unlike the professor on the first day of a semester vocally delineating the terms of engagement if his listener would choose to continue rather than drop the course.

Neither the tactic of quoting the sign nor its particular reference can be taken as casual or adventitious. It exemplifies discontinuities in the contemporary American urban landscape momentous to Eiseley; and his reaction to the sign betokens a conviction that there are privacies

---

15. James Olney, reviewing *All the Strange Hours,* was perhaps the first to argue in print Eiseley's pervasive autobiographical tendency (*New Republic,* November 1, 1975, pp. 30–34).

not to be externalized. In *All the Strange Hours* and his less avowedly or consistently autobiographical writings, there are frequent passages of affecting intimacy. But the intimacy is of individual and private thought. There is not with Eiseley the structure of quasi eavesdropping on two people in private, as with, say, Donne in "The Good Morrow" or "The Canonization." There are indeed Eiseleyan love poems, as in *All the Night Wings;* they are typically of separation and loss and are always hedged by the possibility that they are quasi addresses to an other not merely absent but radically fictive.

Beyond that there are widespread efforts to generalize, to move from the uniquely historical to the designedly emblematic. Partly because "no children watch from the doorway as I write," he explains, he will attempt to "bespeak . . . the autumn years of all men" (*All*, 228). But "our civilization smells of autumn," he had remarked a couple of years before (*Man*, 59), so we may understand the application, the audience potentially addressed, to be very general indeed.[16]

The framing conflict, most general of all because not just of our era or our civilization or even our species, is "the long war of life against its inhospitable environment, a war that has lasted for perhaps three billion years" (*Unex.*, 51). In this "discontinuity beyond all others," inside "has fought invading outside, and . . . won the battle of life." He wrote, "Body controls are normally automatic, but let them once go wrong and outside destroys inside. This is the simplest expression of the war of nature—the endless conflict that engages the microcosm against the macrocosm" (*Inv.*, 48).

Culture, he repeatedly insists, is a kind of secondary, man-made nature, the physical-existential outside made by the thinking animal, the medium in which the war of outside against inside has been doubly represented. But that is to say both doubly and partly. Culture inevitably both exists within nature in some obvious physical senses and contains any notions we form of "nature." Eiseley argues at eloquent length for the historical and provisional quality of any such idea, in the long, midlife essay called "How Natural Is 'Natural'?" (*Fir.*, Chapter 6).

I quote from his later work to illustrate both the persistence of the concern and one direction in which it leads. For my analytic purposes, it

---

16. Peter Heidtmann calls Eiseley an "artist of autumn"—where *of* means both *about* and *making himself verbally from.* See "An Artist of Autumn," *Prairie Schooner*, LXI (1987), 46–56.

is again convenient to reverse the polemical order of Eiseley's book: "Primitive man . . . projected a friendly image upon animals. . . . Man was still existing in close interdependence with his first world, though already he had developed a philosophy, a kind of oracular 'reading' of its nature. Nevertheless he was still inside that world; he had not turned it into an instrument or a mere source of materials" (*Inv.*, 143). This seems to be nodding toward Lévi-Strauss's *The Savage Mind*,[17] but insisting, without knowing Ong's work, on something like the difference between the primitive relation to "environment almost as a single tool" (*Inv.*, 58) and the relation of hyperliterate cultures to environment: "Modern man . . . has come to look upon nature as a thing outside himself—an object to be manipulated or discarded at will. It is his *technology and its vocabulary that makes his primary world.* If, like the primitive, he has *a sacred center,* it is here. Whatever is potential must be enrolled, brought into being at any cost. No other course is conceived as possible. The economic system demands it" (*Inv.*, 59; my emphasis).

Technology and *vocabulary* and *outside* look to spatiality and linearity in Ongian senses characteristic of print culture. A phrase like *sacred center* betokens the Eiseleyan conviction of a latter-day idolatry, a superstitious equivalent of primitive myth. That is a point to consider in a moment, but insofar as culture exists within nature, I must say more about Eiseley's peculiarly ambiguous nature: it is aggressive outside and "queerer than we can suppose," in Haldane's phrase, yet ineluctably "first world" and more validly a "sacred center" than any secondary world.

Clearly Eiseley honors nature most fundamentally as the cradle of life and of the beauties and complexities of life. The relatively simple power and energy of, say, the wind seems too readily adversive to intrigue him as it did Shelley, although a passage from "Adonais" furnishes the title to *The Firmament of Time* and appears as an epigraph to that volume. Nor is he moved by the mountains as brooding "presences" of stability, as Whitehead in *Science and the Modern World* discerned Wordsworth to be.

However grand the geological scene, however uniformitarian (in the geological sense that is opposed to catastrophism), its appeal to Eiseley seems always to have been primarily as a site for life, only secondarily and occasionally as an aesthetic composition. Locally, it can be an over-

17. Owned by Eiseley. For Ong, see especially *Orality and Literacy.*

whelmingly inhospitable site when subject to volcanic eruption, earth-quake, storm, flood, or temperatures destructive to humankind. And looming beyond that is the disturbing problem of the spatially extensive, temporally abrupt "great extinctions." But in what for men must rank as the very large context, nature has been a medium, arena, or theater for life "for perhaps three billion years." It has inexhaustibly fostered life and fruitful transformations of life.[18] These have included the deaths of all individuals and most species, yet not, he insists, with anything re-motely analogous to the rancor or vindictiveness characterizing destruc-tion in culture: "Extinction is an art too great for man, he bungles it / by obscene malice" (*Inno.*, 57). Eiseleyan nature is partly by virtue of its mysteriousness too grand an antagonist for man to triumph over at all. It always transcends any putative "nature," rather as, in a familiar theological formula, God always transcends "God." When man affects to triumph over "nature," the vanquished must include himself, for good or ill.

Culture analogously presents a double menace. In language, materi-ality, economic mechanisms, and institution of power, it became for Eiseley the harsh and imponderable outside to man's caring and vulner-able inside. But because culture is so intrinsically human, to triumph over it, as over nature, tends to entail triumphing over other men. The very science that was part of Eiseley himself, and a loved cause, symbol-ized the conflict. As structure he found it inadequate, as method locally exhilarating but largely disappointing, as partner to technology by no means satisfactorily identifiable as a cornucopia of "better things for better living."

Eiseley's argument that man in twentieth-century Western culture subverts man, culture, and nature can be glimpsed throughout his writ-ings, yet it is an argument hard to anatomize. W. H. Auden saw some-thing of it. But he, like subsequent critics, tended to focus on the writer himself, so pervasive yet so teasingly elusive in the writings. Eiseley's ar-gument might be called double, because it moves compatibly in two modes. The more familiar mode is analytic. Even that is rather—to echo one of his titles—an unexpected universe of discourse, because his ma-terial supports for general assertions range, as he acknowledges, from scientifically conclusive evidence or repeatable demonstration to ra-

---

18. And thereby what Robert Jay Lifton has characterized as "natural immortality," in *The Broken Connection* (New York, 1979), 22–23.

tional probabilities, to interpretable and public bodies of historical data, to personal experience that is arguably typical, to analogy, to allegory and vision. Still, one sees in *The Immense Journey,* in *Darwin's Century,* in *The Firmament of Time,* in *The Man Who Saw Through Time,* and in the chapters and essays of succeeding books the familiar mode of discourse that summarizes findings, weighs evidence, and asserts and supports and concludes. In this, as in the other, call it the existential, mode of argument, there is a deep mistrust of scientism and the myth of science, and disdain for a machinelike exercise of power.

Eiseley's deployment of two modes of argument is not unlike the medieval idea of compatible revelations through the book of Scripture and the book of created nature. It is possible to suppose, without being able to prove it, that Eiseley, the shy, inevitable god of his own universe of discourse, intended such an analogy. Certainly he gave us the argument of his life as well as the argument of his discursive professional knowledge and judgment. Like other men, but more consciously than most, Eiseley was a "man of words." [19] Not surprisingly, then, the argument of his life is the Eiseleyan script of a selective autobiography by no means confined to *All the Strange Hours.*

His existential argument may be sketched with brutal abbreviation. He found as he grew up a world where boys wantonly smash things such as big old turtles (*Night,* 6) or bully smaller, less ordinary boys, where "men beat men," where silences at home might be broken by the hostile words of his (increasingly?) deaf and disturbed mother quarreling with his father. He found an answering hostility and violence in himself. He was too unsparing to deny or sentimentalize it, too kindly to accept it complacently. He represents, rather, a man seeking an alternative life world.

In the first broad alternative term of his existential argument, he found in biological science a nonthreatening community. It appeared manageable, at least at the level of home-built aquarium or field excavation, in contrast to home itself, where his baffled father had commanded him never to "cross" his "not responsible" mother (*All,* 30). In paleontological biology, strictly speaking, there might be no need to kill anything for the increase of knowledge. Yet its world was not dead, to him,

19. The phrase is Carlisle's, in *Loren Eiseley,* 185. He notes Eiseleyan ambivalence about science and its "dimensions of expression"—"almost from the beginning" (pp. 134, 181).

and is not to anyone exercising sympathetic imagination. One sees, to reverse Eliot's formula, the skin upon the skull. And there can be the profoundly encouraging mentor, as he found in Frank Speck, a kind of supplemental father and "magician." Moreover and more generally, the ambience of his work, often separated from urban life and solitary or nearly so, was reassuringly nonchaotic and nonhostile. Eiseley's own words about the writings of Charles Darwin can summarize his sense of the framing satisfactions of science for its practitioners: he speaks of Darwin's work as "one of the greatest scientific achievements of all time: the recovery of the lost history of life, and the demonstration of its total interrelatedness" (*Dar.*, 6).

It is well to remember that Eiseley became a relatively unspecialized scientist, almost a general practitioner. The wide range of science, he knew, revealed with deepening perception and experience less an alternative to the world where "men beat men" than a synthesis, a new articulation of culture embodying transformations of all the old natural and cultural destructions.[20] In his discursive argument, he put the matter that way. In the existential argument conducted as autobiography, he put it as personal crisis.

Perhaps the force and depth of his personal crises are obvious to any reader whose life has not been steadily sunny. But perhaps not. Eiseley's inclination to make the leap of trust and reveal himself is tempered by a courteous man's decorum and reluctance to be aggressively exhibitionistic, a private man's distance, a scarred man's caution. So the exact times in his life and the duration of the crises are a little unclear, the symptoms briefly glossed in phrases about particularly agonistic insomnia and painfully conflicted waking activity. The emphasis falls on a precipitating occasion, but perhaps there were occasions accumulating and gradually decisive.

This factor in the autobiographical argument emerges in at least three different tellings. The last written is the account of a "small death," so called because a small dog died at that time of vast death, World War II. Unlike some that die for science, this dog died in a "needless" experiment, albeit at the hands of a kindly faculty colleague. It died ostensibly as instruction for a class, but for a class that was uninterested, and so it died perhaps mostly as a flourish of the instructor's

20. Carlisle has a bibliography of the normally scientific articles, significantly limited to 1935–1955 (*ibid.*, 188–94).

expertise and authority. When selected from the supply of lab dogs, the dog looked at Eiseley in a way suggesting he recognized his only friend. Eiseley writes that the dog's eyes "haunt me . . . still" (*All*, 145); elsewhere he terms apparently scientific culture "this haunted domain" (*Man*, 57).

Eyes animated the definition and the resolution of his crisis, as these came about. For economy here, I offer a defining passage he wrote earlier, which was, however, about a later time in his life: Christian man as early scientist, he says, drove Pan

> from his hillside. . . . What was gained intellectually was a monotheistic reign of law by a single deity so that man no longer saw distinct and powerful spirits in every tree or running brook. His animal confreres slunk like pariahs soulless from his presence. . . . Finally [science] would, while giving powers to man, turn upon him also the same gaze that had driven the animal forever into the forest. Man, too, would be subject . . . would be exposed. He would know in a new and more relentless fashion his relationship to the rest of life. Yet as the growing crust of his exploitive technology thickened, the more man thought that he could withdraw from or recast nature. . . . Like that of one unfortunate scientist I know—a remorseless experimenter— man's whole face had grown distorted. One eye, one bulging eye, the technological, scientific, eye, was willing to count man as well as nature's creatures in terms of megadeaths. (*Inv.*, 143–44)

Writing a few years earlier, Eiseley had remarked: "As always, there is the apparent break, this rift in nature, before the insight comes. . . . I was the skull . . . the inhumanly stripped skeleton without voice, without hope . . . devoid of pity, because pity implies hope. There was, in this desiccated skull, only an eye like a pharos light, a beacon, a search beam revolving endlessly. . . . Meaning had ceased. There were only the dead skull and the revolving eye. With such an eye, some have said, science looks upon the world" (*Unex.*, 67–68). Those lines introduce an account of the beach at Costabel, in "The Star Thrower," which Eiseley selected for reprinting as the title essay in his last collection and called one of his favorites.[21] The star thrower of that beach is a man who throws washed-up starfish out beyond the breakers, where they have a chance to live. He stands, in Eiseley's meditation, in symbolic contrast to tourists boiling sea creatures in kettles on the beach, to collect shells

21. Loren Eiseley to Lucy Forrester, October 9, 1969, in Eiseley Papers, University of Pennsylvania Archives.

for decoration: nature as Heideggerian standing reserve! An almost Bachelardian revery brings Eiseley, back in his room, a palimpsest of eyes, "beaten bloodshot" eyes of dead animals, finally the troubled eye, in an old photograph, of his deaf, tormented mother.[22]

Costabel, on no map, was an otherworldly place of one strangely kindred man, disparaged and distanced collective tourists, the eye of the man's mother, and the dreadful issue of strife and love. Costabel was and was not Sanibel Island. One may observe a kinship with another landscape, emptied of the commercial traffic and animal husbandry and agriculture we know to have been there, almost allegorically emptied of all but Abraham, Isaac, the knife, and the dreadful issue of obedience or disobedience to a terrible Father.[23] Eiseley acknowledged to the sister of his research assistant that he and his wife had vacationed at Sanibel Island, off the Florida coast. With autobiographical evasiveness but thematically apposite generosity, he later answered an inquirer that "I occasionally make use, for literary and other purposes, of created names. It frequently serves stylistic purposes and lends a timelessness which can sometimes not be maintained around a particular geographic point."[24] In short, he expunged the trivially historical and most of the immediately historical. He omitted from the story his wife, who is disregarded like Abraham's wife when the issue is the sacrifice of a third member of the family.

"Love not the world": Eiseley cites the biblical injunction, then recalls answering, "But I *do* love the world . . . the lost ones, the failures of the world." He adds that "it was like the renunciation of my scientific heritage" (*Unex.*, 86).[25] Clearly enough, this is a stark and intimate account of something that almost reduced life to nullity, and of what reso-

22. A somewhat more vivid eye in his mind and prose than in the pictures in *The Lost Notebooks of Loren Eiseley*, ed. Kenneth Heuer (Boston, 1987).

23. Erich Auerbach has pointed this out in *Mimesis: The Representation of Reality in Western Literature* (Princeton, 1953), Chapter 1.

24. Loren Eiseley to Eloise Elsea, June 10, 1968, Loren Eiseley to William Embler, May 28, 1970, both in Eiseley Papers.

25. See also Leslie E. Gerber and Margaret McFadden, who see the resolution as Eiseley's recognition of not merely hating his mother but also, and after all, loving her (*Loren Eiseley* [New York, 1983], 119–20). But she was not the only member of the Eiseley family or of the human (and still larger) family he both loved and hated, though she was a crucial test. Carlisle emphasizes that Eiseley dropped the maternal C. from his title-page name and blamed her even for his own childlessness (*Loren Eiseley*, 150–51, 5).

lution permitted life to go on. It comes from a man who increasingly described himself as a typical American. It gains confirmation from an analogous account written several years earlier, which he likewise selected to be republished in *The Star Thrower*.

"How Natural Is 'Natural'?" is the climactic essay in *The Firmament of Time* and was originally developed for a series of lectures at the University of Cincinnati. Its discursive context and quality make it useful here additionally as a bridge to the quasi-formal mode of the Eiseleyan argument:

> On the morning of which I want to speak, I was surfeited with the smell of mortality and tired of the years I had spent in archaeological dustbins. . . . It was time, I thought, to face up to what was in my mind—to the dust and the broken teeth and the spilled chemicals of life seeping away into the sand. It was time I admitted that life was of the earth, earthy, and could be turned into a piece of wretched tar. . . . It was time I looked upon the world without spectacles and saw love and pride and beauty dissolve into effervescing juices. I could be an empiricist with the best of them. I would be deceived by no more music. . . . "Why did we live?" There was no answer I could hear. (*Fir.*, 160–61, 165)

Eiseley drives the essay to an existential answer, though he draws the answer from another field trip he made either earlier or later in biographical time. On it he saw a blacksnake and hen pheasant entangled in combat that promised to be fatal to one or both. He separated them and carried the snake out of range, coiled around his arm. He understood that the pheasant struggled for a pheasant future with hatchling brood, and that the snake struggled for a snake future with pheasant-egg lunch. After meditating on this composition of place,[26] Eiseley concludes that in pacifically separating them he himself "struggled . . . for a greater . . . more comprehensive version of myself" (*Fir.*, 176). Here is the recurrence of an Eiseley-family emblem, as when, a frightened child, he separated quarreling parents (*All*, 24; for the snake and pheasant, p. 235). Or did he think of the Laocoon?

In any case, the struggle admits of successes, and every Eiseleyan collection furnishes exempla. But the uneasiness in Zion, the recurring, vi-

---

26. My phrase deliberately echoes Ignatius of Loyola, as discussed in Louis Martz's *The Poetry of Meditation* (New Haven, 1954), and glancingly in *The Paradise Within* (New Haven, 1964), the latter of which Eiseley owned and acknowledges in his bibliography for *The Invisible Pyramid*.

tal trips into the desert, the diet, so to speak, of pheasant eggs and As-
teroidea, like locusts and wild honey, yield counsel of repentance and
heed for the wrath at hand.

Eiseley's analytic mode of argument when as prophet and natural
philosopher he becomes a formal critic of science's objectifying gaze
conforms to his existential insight. Another essay in *The Firmament of
Time* appears there in virtually the same words that it appears as "The
Ethic of the Group," his contribution to a symposium at the University
of Pennsylvania.[27] Eiseley was a practicing scientist to the last. The enemy is not Baconian science in its life-enhancing thrust against one or
another constraining context of myth, nor technological aids to health
or community. As he celebrates Darwin for disclosing the interrelatedness of life, so he celebrates Bacon for seeing beyond the "circle of mechanical technology," for seeing new learning as means "to bring an enlightened life" (*Man*, 14; *Inv.*, 69). Rather, the enemy is a reductive and
institutionalized science and its offspring technologies, science itself
gone mythic, triumphant, unloving, and therefore oppressive, yielding,
as someone has said, not the rule of law but the law of rule. The enemy
is science that, answering to Eiseley's language of attenuation and carry-
ing-off, might be considered *ablative:*

> Science as we know it has two basic types of practitioners. One is the edu-
> cated man who still has a controlled sense of wonder before the universal
> mystery, whether it hides in a snail's eye or within the light that impinges on
> that delicate organ. The second kind of observer is the extreme reductionist
> who is so busy stripping things apart that the tremendous mystery has been
> reduced to a trifle, to intangibles not worth troubling one's head about. The
> world of the secondary qualities—color, sound, thought—is reduced to illu-
> sion. The *only* true reality becomes the chill void of everstreaming particles.
> ("Science and the Sense of the Holy," in *Star*, 151)

He writes of "controlled sense," that is, the sense of wonder of someone
who is tough-mindedly aware of knowledge and the constraints on
knowledge and is not simply gushy. He writes of the "chill void," a fascinating nightmare for a man who repeatedly sortied from enclosed
hearth, always to return, and who sometimes imagined himself "on the
altiplano" in an Ice Age scene, huddled with his dog.

27. Loren Eiseley, "The Ethic of the Group," in *Social Control in a Free Society*, ed.
Robert E. Spiller (Philadelphia, 1960), 15–38.

In "How Human Is Man?" Eiseley insists on the historicity of science: "Bacon's world to explore opens to infinity, but it is the world of the outside. Man's whole attention is shifted outward. Even if he looks within, it is largely with the eye of science. . . . Concentrated upon things . . . it hungers for infinity. Outward in that infinity lies the [Paradisal] Garden the sixteenth-century voyagers did not find. We no longer call it the Garden [but], vaguely, 'progress.' . . . We have abandoned the past without realizing that without the past the pursued future has no meaning, that it leads . . . to . . . dehumanized man" (*Fir.*, 130). The "technological revolution," he continues, "first . . . has brought a social environment altering so rapidly with technological change that personal adjustments to it are frequently not viable. . . . Second, much of man's attention is directed exteriorly upon machines" (*Fir.*, 133–34). "Third," he continues, "this outward projection of attention . . . upon . . . power torn from nature" now too easily lacks "individual conscience." Conscience is an allegedly secondary quality. Part of the problem is the unpredictable way in which the gigantic power of current science and technology "brought into the human domain . . . partakes of human freedom. It is no longer safely *within* nature; it has become violent, sharing in human ambivalence and moral uncertainty" (*Fir.*, 135). Approximations to one or another of these three points have been made by assorted twentieth-century observers. Eiseley's own amalgamation and follow-up belong with the formulations of Tate and Mumford: "sharing in human ambivalence and uncertainty" evidently is a way of Eiseley's for recognizing "objective" scientific laws and their applications as *cultural* artifacts. And unlike allegedly value-free social scientists, he would differentiate good cultures and bad.

Eiseley seems to have come in subsequent years to the argument that I take to be peculiarly typical of him: that of linking the three consequences of the technological revolution that he has enumerated with his grand preoccupation, the human sense of *time*. Technological man, scientistic man, is anticipatory analytical man, enraptured by novelty: "In simple terms, the rise of a scientific society means a society of constant expectations directed toward the oncoming future. What we have is always second best, what we expect to have is 'progress.' What we seek, in the end, is Utopia. In the endless pursuit of the future we have ended by engaging to destroy the present" (*Inv.*, 105). In *The Invisible Pyramid*, his "moon book," he does not dwell on the vulnerability or thinness of

the present as a concomitant function of losing the past. But that idea
pervades his work and might be summarized as, We do not know *who*
we are if we do not know *how* we were. More grandly and urgently
than Van Wyck Brooks's or Carl Becker's search for a "usable past" in
terms we might have thought generous enough, this is an attempted recovery of an enormous past to enliven a present, which need not lead to
entropy or a mechanically extrapolated future.[28] Eiseley would avoid,
and have us avoid, the present-to-future compulsiveness of anticipatory
man, as we would avoid the equally tendentious past-to-present sallies
of medieval or Renaissance typologists.

It is not necessarily conclusive that *we have* more knowledge about
the past than ever before. *Have?* We know, say, more about the classical
world than Alexander did, and more about the Stone Age than any
nineteenth-century person did. *We?* But anticipatory man is disconnected from the past by the ready combination of instrumentalist attitude and the cemeterylike physical nature of libraries, still more the
warehouselike nature of data banks (as ineptly called memory banks as
the steam locomotive was once called the iron horse). Given that combination of obstacles, the past becomes not presences giving our now
some of the thickness—the readiness, the kindred richness and consequentiality—of then, but a congeries of events, instrumentalities in
space with no presence beyond occasional grist value for a utilitarian
mill. The semi-quasi-Derridian poststructuralist argument against presence—the protest against privileging voice, and the denial that there can
be a god echoed or represented by true voicings—this argument, which
may be anchored in faith and may foster some valuable consequences
but which privileges sight, can serve well enough as an active report on
the technologistic attitude that looks upon language as a lifeless pattern
on the page, as the substance and structure of differences in space. That
is, presence is absent nonetheless for having been excluded by reason
of ideological or psychological impulsions, whether neurotic powermongering or some other.

The poetically inclusive and affectionate sensibility would see loss as
well as gain in Pan's displacement from the hillside: less *panic* but also
less ecological respect, say. That sensibility has its own liabilities.
Eiseley identifies an escapism, a kind of historian's rapture of the deeps,

28. See also Angyal, *Loren Eiseley,* 38.

as when he himself was working on *Darwin's Century:* "The scholar who descends into the catacombs of the past is endangered; he may lose his way, . . . the future . . . unconsciously abandoned" (*All,* 183–84).

While acknowledging the usual regularity of nature and even of the man-made nature that is culture, regularity against which at least the rate of current changefulness is *unusual* (see above, p. 210), Eiseley goes beyond the commonplace that myth always involves repetition whereas history does not repeat itself. He regards history as so much the abode and medium of the unique that to slight either history or uniqueness is to slight the other. At the same time, history is the only locus for an "eternity"—which, probably through a debt to Kierkegaard, means to Eiseley transcendence of a reductively mythicized future. Man without history is a "fabricator of illusions," he averred as early as 1958 (*Dar.,* 28). It is a matter of groundedness, of appropriately unpredictable complexity. *Being* for him in its imagination-testing temporal reach and mysterious biological imponderability is, as it was for Rilke, *round*. In modern science, as he construes it, the ready and present myth, the privileged illusion, is that of the machine. Eiseley underscores this through a glance back at the eighteenth-century geologist James Hutton, acknowledging in somewhat wry appreciation that "his world machine sounds at times like a heat engine" (*Fir.,* 28). But he is not wry in referring to the "state-manipulated machine" of dogmatic ideology (*Man,* 100–102) against which Bacon struggled. Today the struggle is with a religious or mythic commitment to institutionalized externalities, regularities, and transiences.[29]

I mean by *myth*—and Eiseley means by an institutionalized scientism and technologism, the pharos light and anticipatory man—a systematizing, totalizing mental pattern for dealing with semiconscious assumptions, conflicts, desires, and fears and integrating problematical experience with them, a pattern for dealing that permits social action (*Imm.,* 199).[30] Hence, myth is doubly reductive: it *reduces* existential anxieties by *reducing* the meaningfulness that goes with undecidability.[31] Eiseley complained that normal "science seeks essentially to

29. Gerber and McFadden speak of a "fatal Faustian bargain with machines" (*Loren Eiseley,* 83).

30. My thinking on myth and the twentieth-century American prophetic impulse has been helped at every stage by Herbert N. Schneidau's *Sacred Discontent: The Bible and Western Tradition* (Baton Rouge, 1976), esp. 7, 10, 42, 263.

31. Gerber and McFadden helpfully underscore Eiseley's attention (*Dar.,* 347–52) to

naturalize man in the structure of predictable law and conformity" (*Night*, 138; *Man*, 104). Worse, "to those who have substituted authoritarian science for authoritarian religion, individual thought is worthless unless it is symbol [sign, we would say] for a reality which can be seen, tasted, felt, or thought about by everyone else. Such men . . . reject the world of the personal, the happy world of open, playful, or aspiring thought" (*Night*, 139; *Man*, 104–105).

Authoritarian science's myth, then, must be recognized as banefully exteriorizing, ahistorical (and with that, disablingly regularizing), and fetishistic about predictability. Eiseley, in opposition, wants a horizon open to what? To the unexpected. "Playful"? Buoyant in knowingly *limited* control. "Aspiring"? Acknowledging transcendence and mystery.[32] Like Mumford, he sees the scientistic myth as not only machine-like but increasingly global in the individual personal sense and the political sense of the corporately personal: "The 'within' and 'without' are in some fashion intermingled. . . . It is obvious that the whole of western ethic, whether Russian or American, is undergoing change, and that the change is increasingly toward conformity in exterior observance and, at the same time, toward confusion and uncertainty in deep personal relations" (*Fir.*, 120–21). Argument goes on at large over whether or not personal relations are thinned by ritualization in an oral-formulaic culture. But the surmise is that few in contemporary culture have not felt at least occasional conflict between home concerns and the relentlessly linear profit-and-loss concerns of the office.

The associated technologism should be seen, he insists, as elaborating and reinforcing tendencies to external conformity with its neurotically mythic regard for the predictable, repeatable, replaceable—or, in different words, the commoditized and instrumentalized. "The economic system demands it," he remarks (see above, p. 202). Although he does not elaborate, the profit machine parallels in his thought the "state-manipulated machine" (see above, p. 212). A letter he wrote to an executive after a blizzard remarks, "I do not question that the Financial problems [of railroad passenger service] are difficult ones, but there are

---

the brain as the organ of *indetermination*, and they note that he got the "term from Henri Bergson" (*Loren Eiseley*, 83).

32. Gerber and McFadden rightly draw attention to an early focus by Eiseley (*Dar.*, 336) on the mystery of life's tendency to collect itself in ordered wholes (*Loren Eiseley*, 96). After all, why has entropy not always already forestalled evolution?

some unlovely aspects of big capitalism and its abrogation of social responsibilities which show through at moments like this." [33] Here, as
elsewhere, Eiseley's attention tends toward the range of the myth more
than toward particular corollaries or elements such as power- or profit-
mongering or commoditizing. And his concern, therapeutically perhaps,
tends toward the psychodynamics of the myth's operation, that is, toward the psychic rewards for those who embrace it and the alternative
rewards for those who find another path. He observes that Shakespeare
"saw at the dawn of the scientific age what was to be the darkest problem of man: his conception of himself" (*Night,* 55).

As for resistance to the myth of science, "authoritarian science," I
want only to indicate here something of the direction of his critique
as he moves from its terms to his own, which I shall give separate
treatment.

Perhaps the most obvious allegation he makes—obvious because recurrent in his writings and well established in the Western tradition—is
that suppositions of machinelike efficiency are fraught with a temporal
parochialism. When a machine will break down may be unpredictable,
and that it will break down at any particular moment may be too unlikely to be significant for some human purposes. But that it will break
down sometime is certain, and no one—not the Robinson Jeffers he
honored, nor any archaeologist—has a more vivid sense than he of
technics rusting and crumbling, weeds thrusting through Philadelphia's
Market Street, and the winged and four-footed inheriting the ruins. Machines are always inefficient in any but the shortest run, fit only for a
twitch of time. And they tend to reduce time to a twitch: as Ernst Cassirer has pointed out, science asks primarily what *is* (or what *was,* construed in atemporal terms), technology what can be, construed in func-
tionally, rather than imaginatively, extrapolative terms. [34] Technology,
Mumford seems to object, seeks for or pretends to complete disclosed-
ness and availability, Heidegger's *zuhandensein.*

Moreover, there are ironic ambiguities that Eiseley is not slow to
point out in our fondness for the machinelike replaceability of human
parts, balance-sheet reductivism, and the like: "We must consider the
possibility that we do not know the real nature of our kind. Perhaps

33. Loren Eiseley to Charles Aring, January 11, 1970, in Eiseley Papers.
34. John Michael Krois, "Ernst Cassirer's Theory of Technology and Its Import for
Social Philosophy," *Research in Philosophy and Technology,* V (1982), 210.

Homo sapiens, the wise, is himself only a mechanism in a parasitic cycle, an instrument for the transference, ultimately, of a more invulnerable and heartless version of himself. Or, again, the dark may bring him wisdom" (*Inv.,* 54–55). The questioning is prophetically radical.

Homo sapiens going into the dark, as unnumbered individuals and even species have done and will do, suggests animal-like organicism. But no one has turned a more unsparing eye than Eiseley has on woolly-minded organicism:

> I have said [civilization] is born like an animal and so, in a sense, it is. But an animal is whole. The secret tides of its body balance and sustain it until death. They draw it to its destiny. The great cultures, by contrast, have no final homeostatic feedback like that of the organism. They appear to have no destiny unless it is that of the slime mold's destiny to spore and depart. Too often they grow like a malignancy, in one direction only. The Maya had calculated the drifting eons like gods but they did not devise a single wheeled vehicle. (*Inv.,* 132)

Time in its limitlessness and history in its rich uniqueness are the mediums of escape from a myth of the machine on one side and a myth of animality or organicism on the other.[35] But time and history must be apprehended with an extraordinary largeness of view, and animated with unusual generosity:

> Like Whitman, like W. H. Hudson, like Thoreau, Ishmael, the wanderer, has noted more of nature and his fellow men than has the headstrong pursuer of the white whale, whether "agent" or "principal" within the Universe. The tale is not of science, but it symbolizes on a gigantic canvas the struggle between two ways of looking at the universe: the magnification of the poet's mind attempting to see all, while disturbing as little as possible, as opposed to the plunging fury of Ahab with his cry, "Strike, strike through the mask, whatever it may cost in lives and suffering." (*Star,* 160)

In an essay in the *American Scholar* in 1964, Eiseley argued against C. P. Snow and any who hold to, as the essay's title has it, "the illusion of the two cultures." That essay reappears in *The Star Thrower,* joined with one from the *American Scientist* from 1963, in which the crux is his assertion, with a genuflection toward Whitehead, that "all responsible decisions are acts of compassion and disinterest; they exist within

35. See also Gerber and McFadden, *Loren Eiseley,* 72.

time and history but they are also outside of it, unique and individual and, because individual, spiritually free" (*Star*, 218).

That can serve as conjunction and conclusion for the two modes of Eiseleyan argument, the existential and the discursive. It suggests the track they jointly follow. It suggests the larger case: each of Eiseley's essays, beyond the occasional local flaws of self-interest or indulgence, is essentially unique, individual, disinterested, and compassionate, and is by free choice constituted as less an act of power than a gift.

Is there not something extraordinary about the tone in which Eiseley enunciates his teacherly, prophetic addresses? It seems to indicate an idiosyncratic posture toward the world. Is it prophet as fugitive familiar? Auden, sounding a little like Samuel Johnson finding always some melancholy in Milton's mirth, supposed that Eiseley was a melancholic. To determine the biographical appositeness of Auden's idea would be quite a different project, would involve weighing the abundant testimony of surviving friends that he was often a mirthful host or guest or coworker, might entail isolating and scrutinizing the moments of dry humor in the writings (not unlike boiling champagne, that tactic), would involve inconclusive psychobiography that I am scarcely equipped to attempt. Others, reading Eiseley's essays, may think more than once of Walter Benjamin's "story teller . . . who could let the wick of his life be consumed completely by the gentle flame of his story." [36] But the complex doubleness, the mixed sense of intimacy and estrangement, conveyed by his language seems so important a part of his transaction with his reader that one fruitfully considers features even if the whole resists formulation in a phrase.

Eiseley himself felt consciously and keenly a sort of dismaying cultural estrangement. In *Unexpected Universe*, he invokes wandering Od-

---

36. See W. H. Auden's review article in *New Yorker*, February 21, 1971 (reprinted in part as the introduction to *The Star Thrower*. The reference is to Walter Benjamin's "The Storyteller," in *Illuminations*, trans. Harry Zohn (New York, 1969), 108–109. Compare Eiseley's tone and posture in the scientific articles of the forties: "watchful, critical, suspicious, and somewhat contentious . . . very careful, and he exposes himself very little," as Carlisle has well put it (*Loren Eiseley*, 129).

ysseus as a hero and emblem of Western culture betokening that "there is nothing worse for mortal man than wandering" (repeated as epigraph of "Days of a Drifter," in *All*). The point is that "there is a sense in which the experimental method of science might be said merely to have widened the area of man's homelessness" (*Unex.*, 48).

The feeling and conviction on Eiseley's part extended from the more cerebrally cultural to the more practically and personally social. He answered a letter telling him of the death of Letta Mae Clark, who had greatly encouraged him when he was her high-school English student in Lincoln, Nebraska:

> With her passing, indeed even when I learned she moved to Sacramento, I could feel my own ties with Lincoln slowly slipping away. When one has lived a long time in the great cities of the East one feels quite homeless as one approaches retirement. The population is so highly mobile that as one's friends vanish around one it becomes more and more of a problem what one should do in retirement. One cannot turn back to childhood, and most of the friends I once had in Lincoln are either dead or long since dispersed to unknown parts of the country. Very few close relatives survive, and although we would like to leave the Philadelphia area we would scarcely know where to turn.[37]

"Great cities of the East" sounds almost like the phrase of a schoolboy fantasizing skylines of Boston–New York–Philadelphia–Washington from the "sunflower forests" of Salt Creek at the edge of Lincoln. Yet he got around those cities adeptly—got around them institutionally, socially, practically, almost routinely. On the other hand, what of the reflection that "most of the friends . . . are either dead or long since dispersed to unknown parts . . . We would scarcely know where to turn"? It is a fact that financially pinched universities—and there is no other kind—are awkwardly unable to do much for their own retired faculty, much less for the retired faculty of other universities, even when the persons are distinguished. But it is not rationally conceivable that the University of Nebraska would have been inhospitable to so distinguished an alumnus, one to whom it had awarded a D.Litt. in 1960. There was and is the Nebraska Academy of Science. And although some friends had indeed died by 1973 and others would have proved difficult or even im-

37. Loren Eiseley to Mrs. Arnold Baragar, November 16, 1973, in the Heritage Room Collection of the Bennett Martin Public Library, Lincoln, Nebr. Quoted by permission of Lincoln City Libraries, Mrs. Baragar, and the Estate of Loren Eiseley.

possible to trace, Eiseley had friends in the Lincoln-Omaha area with whom he had been in affectionate and fairly regular communication, friends for whom he is to this day a significant presence. Unsurprisingly, that is true in Philadelphia as well, although an important companionship was dissolved when Wright Morris left for California and his wife remained; they had for him been a particularly congenial couple.[38] The truth seems to lie deeper, at a social-metaphysical level perhaps parallel to the erotic-metaphysical level at which Eliot wrote of Donne that "no contact possible to flesh / Allayed the fever of the bone."

Anyone reading *The Man Who Saw Through Time* after reading *All the Strange Hours* is likely to be struck by a recurring motif: a kind of extended metaphor that Renaissance rhetoricians would have called allegory. Somewhat as Izaak Walton saw Saint Augustine's life as a metaphor for Donne's, Eiseley apparently saw Francis Bacon's working life as a metaphor for his. The sense of being a fugitive is one aspect of it: "Bacon . . . condemned to spend his life in a world of . . . court intrigue . . . chose to have his actual being in a far different one of his own making. . . . [He] died broken and forlorn in an age he never truly inhabited" (*Man,* 51).

Of course the Eiseleys' world in Philadelphia was increasingly a world of their own making, "childless and my destiny not bound to my kind" (*Inv.,* 133). Nearer and nearer the end, Eiseley seems to have appreciated more clearly, and certainly he was more explicit about, what his kind had been. A very late poem proposes an allegorical gloss on his fugitive status even as it dramatizes other things, some of them possibly exceeding his own awareness. He called it "A Hider's World," and it goes, in its most relevant part:

> Once in my callous evil youth I saw a bittern
> take two steps in the reeds of a swamp thus ceasing
> to be a reed. Then it straightened again, pointed up its bill
>   and I
> lost it or pretended to. My companion did not see it,   5
> it was a reed again and I chose to lose it.
>   Since then
> I am bittern-minded, try to keep my own place in the reeds
>
> . . . .

38. See Wright Morris, *A Cloak of Light: Writing My Life* (New York, 1985). Morris was presumably the friend in *The Unexpected Universe,* 64–66.

I

have survived the disguise of a teacher,   20
dusted my clothing with chalk, spoken
to the unlistening, but for what, I want to know
now that it is ending. Why does a bittern stand
so successfully on one leg? Is this the purpose
he was formed for and as for me, dusted with chalk,   25
eyes not to be seen on a dark night, what was the
                                    purpose
                                    engendered
in me? To love, and conceal it all of my life
                        like the bittern   30
trying to be a reed? We were necessary failures, bird,
necessary to keep something alive that the time
                        is not ripe for.

. . . .

I lost my youth and laid the rifle
quietly aside for the sake of hiders. That, I think,
                has been my purpose, a hider's world.

                                    (*Ano.*, 45–46)

This ancient mariner did not slay the albatross/bittern nor show himself anything of a "callous evil youth," whatever his companion may have been. Those words manifest a sort of ironic exaggeration quite uncommon with Eiseley, as if they were an apotropaic charm warding off rage he elsewhere acknowledged remorsefully in himself. Although from the moment of choice he has been "bittern-minded," he has bred "other reed actors" only insofar as his poetic and expository pedagogy may have done so. The poem, one notes, adverts neither to the biological fact he treats elsewhere, nor to the cultural possibility—even seems to deny the latter, since the teacher is defined as a quasi teacher and his audience as unlistening. This last comes from someone who modestly admitted to "some followers" (*All*, 131) as a junior faculty member, who had received a citation in 1962 from the Pennsylvania Department of Public Instruction, and whose books, which had won awards and achieved heavy sales, were selling briskly even as he wrote the poem. "No contact possible to flesh . . ."

No more is he a murderous mariner than he is a tattered coat upon a stick who would sing for tatters in his mortal dress. To make and to foster a "hider's world" may be a spiritual enterprise, and he elsewhere reflects as if in surprise that his whole life "has been a religious pilgrim-age," but clearly his focus is the natural world and his medium for self-realization secular. And he, like the bittern (in an irony *not* exagger-ated), evidently felt compelled intermittently to make self-revealing moves. That the mortal dress and voice were until the last not tattered but imposingly fine may even have been a positive factor in emboldening that self-revelation, not necessarily nor exclusively through vanity but in affectionate analogy to a father who declaimed Shakespearean verse.

The moves by which he revealed himself are for him in any case both individual retreats and representative hearkenings, not only from, but toward: "Something still touches me from that vanished [glaciated] world as remote from us in years as an earth rocket would be from Al-pha Centauri. Certainly Cocteau spoke the truth: to add to all the cos-mic prisons that surround us there is the prison of the golden light that changes in the head of man—the light that cries to memory out of van-ished worlds" (*Inv.*, 127–28). The plurality *worlds* is arresting. Eise-ley's particular and often-acknowledged hearkening—toward the city-less, nearly technologyless world of man and animal in epic struggle against the cold—that hearkening may appall some readers at almost every level of consciousness. What is the reward to be found imagina-tively close to the ice? It contrasts as if defensively with what Auden called the "warm nude ages of ancestral poise." And those ages are equally with the glaciations verified in paleoclimatology, equally plaus-ible in candidacy for any collective unconscious there may be, and equally assimilable to interpretation in terms of regressively desired infantile re-lation with a nurturing mother. Or the glaciated world may be a projec-tion of the inner feeling of being chilled. But clearly the confession of idiosyncratic kinship across time is an atechnic handclasp across social space.

Eiseley owned and seems to have known Bruno Bettelheim's *Sym-bolic Wounds*, which expresses deep skepticism about the Freudian pos-tulate of "memory traces," especially when clinical observation or an-thropological field observation suggests some different hypothesis. Eiseley was an evolutionist who rejected Lamarckian and other supposi-tions of the inheritance of acquired characteristics. But he noted and commented on insect behavior so compound and complex that it can scarcely be conceived to have evolved by degrees. He describes a faithful dog, Wolf (dedicatee of *The Unexpected Universe*), just roused from sleep, fighting with him, the beloved master, for a *fossil* bone. No doubt

Wolf could also have been seen, like other dogs, to turn around and around before lying down on a residential floor, as if hollowing a space in archaic grassland.[39] Whether Eiseley's observations moved him to think of a collective unconscious in the Jungian vein or of creative memory in the Augustinian is hard to say. But clearly part of "this protest of a fugitive" (*Wings*, 47) is the belief in mysterious resources that in kindred minds oppose the functional, extrapolative linearity of anticipatory man.

A pattern of conflictual alternation between the attitudes of current pseudopracticality and the saving and troubling deliverances of memory is one he finds within himself: "I put the matter out of my mind, *as I always do*, but I dozed and it came back" (*Unex.*, 28; my emphasis).

Joan Bennett once argued that we read the quality of the beloved and of the lover's emotion in a Donnean love poem not in any praiseful catalog of her beauties but rather in the lady's effect on the person whose voice we hear in the poem. The structurally analogous argument needs to be made with regard to Eiseley, whose prophetic voice speaks with idiosyncratically careful familiarity in the published writings, speaks variously in the letters, speaks to Eiseley himself with a degree of ellipsis in the notebooks, and spoke not dissimilarly in live voice to friends.

Investigation quickly reveals a body of friends, some of very long standing and typically of profound loyalty. One veteran of the South Party was listening, when I visited him, to a Library of Congress recording of *The Unexpected Universe*, fifty years after fieldwork with Eiseley. Wright Morris celebrates the friendship in his autobiography, recalling that the two couples were often together. A chance conversation I had with a young M.D. on a westbound plane five years after Eiseley's death disclosed that she considered him her finest undergraduate teacher, in effect such a hidden teacher as he himself celebrated in Frank Speck ("The Last Magician," in *Inv.*).

That essay is an extended instance of a recurring dramatization: flight from anticipatory man yields connection, then yields familiar presence—from a professional convention into the confidence of Frank Speck; from whatever scramble of professional duties into the confidence of the marine engineer "Tim Riley"; from a seemingly over-ritualized Christmas party into the confidence of a fugitive but commu-

39. It is possible to say, "no doubt," because Wolf is so considerably fictionalized. See Angyal, *Loren Eiseley*, 81–82.

nicative cat; from museum-sponsored field-camp routine into something approaching the confidence of a "Neanderthal" girl, about whom there will be more in the next chapter.

One member of that metacarpal-digging field party called the girl retarded.[40] With regard to her, as to the others whose presence he has reached, Eiseley gives a stereoptic view: in this case, the normal, marginalizing view, yoked to his revisionary view of the remarkable unevenness of capabilities, the jagged profile of what we crudely call intelligence in anyone. Eiseley's habit of fleeing the mechanistically predictable and totalizing put him in readiness to expect the rare or unique and to honor it. So appreciatively does he focus his own and our attention on the extraordinary that we could easily overlook the pattern in his pilgrimage and the significance of his success. It is true but not enough to say, "The paradox of [Jeffers'] daemon: 'to escape and not to love; to love and not escape' . . . Eiseley's demon, too."[41]

His function as prophet identifying the idol worship crushing us may obscure the peculiar nature of his success, and certainly his own published chronicles of rejection do. Merely to sample the range is to see a hider's habit at work. What he says can be buoyantly near-jocular, as when he writes, explicitly enough, in his often metaphorically autobiographical book on Bacon that his reflections on Bacon occasioned "sufficient controversy to make me wonder whether it was I who was threatened with the Tower and whether Parliament was in full cry upon my own derelictions" (*Man*, 13).[42] It can be a sober interpretation of generic rejection:

> The spore bearers, once they have reached the departure stage, are impatient of any but acceptable prophets—prophets, that is, of the swarming time. These are the men who uncritically proclaim our powers over the cosmic prison and who dangle before us ill-assorted keys to the gate. (*Inv.*, 124)

Or it can be a dry report of a sweaty institutional rejection:

---

40. Emery Blue, interview with author, Lincoln, Nebr., July 1, 1985.

41. Carlisle, *Loren Eiseley*, 118. Carlisle is quoting from Eiseley's foreword to *Not Man Apart: Lines from Robinson Jeffers, with Photographs of the Big Sur Coast*, ed. David Brower (San Francisco, 1965).

42. Many places in *The Man Who Saw Through Time* suggest metaphoric identification, but see especially pp. 33, 46–47, 50, 51, 56, 61, 81. Compare Angyal, *Loren Eiseley*, 72.

I believe I may say that I resurrected [Edward Blyth] sufficiently that considerable energies at Cambridge and elsewhere have been devoted to laying his ghost, not with entirely satisfactory results. . . . I knew that my curiosity was going to bring down on me the wrath of those who regarded Darwin as a sacred fetish, a scientific saint who had appeared with the tablets directly after a mysterious world voyage, an Odyssean adventure. (*All,* 191–93)

Or it can be dignified yet pained reports of personal rejection, as by "Jimmy Dawes,"[43] or of professional rancor, as at a convention cocktail party, or of campus curses at his Franklin Professorship (*All,* 159–62, 181, 202–203).

Few who have observed such behavior as he reports will call Ortega's words on the general subject overinterpretation:

[Rancor] is the imaginary suppression of the person whom we cannot actually suppress by our own efforts. The one towards whom we feel resentment bears in our imagination the livid semblance of a corpse: in our minds, we have killed him, annihilated him. Later, when we find him actually sound and unconcerned in reality, he seems to us a refractory corpse, stronger than ourselves, whose very existence is an element of mockery, of disdain towards our weakness.[44]

Some years ago I was introduced to a certain scholar, a man who had formerly been at the University of Pennsylvania. "You must have been there when Loren Eiseley was provost," I exclaimed. "A *poet,*" he snorted. In saying of one such would-be executioner, "I knew what animus had always consumed him, for reasons unknown to me" (*All,* 181), Eiseley is modestly acknowledging an Ortegan discrimination—perhaps fostered by the very text cited above—of causes, and modestly disclaiming knowledge of final cause somewhat in a Pauline or Augustinian sense of the "mystery of iniquity."

More certainly, his words and behavior clearly evidence awareness in these terms of the dilemma within himself. He had wanted to kill a murderously threatening railway brakeman but apparently instead first withdrew emotionally in stoic endurance and then physically fled (*All,* 7–8). He apparently wanted to kill an image of his dangerously threat-

43. Anyone seeking the actual name might note that *Dawes* names a northwest Nebraska county in a general region of fossil hunting, whereas monosyllabic Gage is nearby and the city of Lincoln is in Lancaster County.

44. José Ortega y Gasset, *Meditations on Quixote* (New York, 1961), 35.

ening mother. But even the faintly symbolic killing by running away (*All*, 32) was rendered onerous for him by the injunction of his beloved Shakespeare-declaiming father: "Do not cross her"; she is "not responsible" (*All*, 30). Compare "Taint not thy mind. . . . Leave her to Heaven." The near–double bind, so akin to Hamlet's, was almost as fatal as the injunction to the prince.

Like one of the biological "failures" he identified with, however, he found a narrow rent in the screen of his existence, a difficult way into a livable niche. He fled for years, fled from the deathbed of his father, his mother abusive even there.[45] He did not in the literal sense kill either the threatening mother or (quite) himself. He biologically killed his genealogy, resurrected some of it as print, and adopted in life and in print the boundless potentialities of the living world:

> Her whole paranoid existence from the time of my childhood had been spent in the deliberate distortion and exploitation of the world about her. Across my brain were scars which had left me walking under the street lamps of unnumbered nights. I had heard her speak words to my father on his deathbed that had left me circling the peripheries of a continent to escape her always constant presence. Because of her, in ways impossible to retrace, I would die childless. Today, with such surety as genetics can offer, I know that the chances I would have run would have been no more than any man's chances, that the mad Shepards whose blood I carried may have had less to do with my mother's condition than her lifelong deafness. But she, and the whisperings in that old Victorian house of my aunt's, had done their work. I would run no gamble with the Shepard line. I would mark their last earthly appearance. Figments of fantasy I know them now to be, but thanks to my mother and her morbid kin they destroyed their own succession in the child who turned away. (*All*, 223)

The explanation is overdetermined, no doubt: fear of madness, economic prudence, public-spirited determination not to reproduce madness,[46] vengefulness as hinted in the ambiguously active and passive

---

45. *Even there?* Of course one understands something of a deaf woman's feelings of helplessness, abandonment, and consequent anger in the presence of the corpse, prospective insurance dollars, and a son who may have seemed to *her*, even if to no one else in the world, privileged and carefree.

46. My friend Dr. Stanley Miller first suggested to me that Eiseley's stated reason scarcely accounts for the matter. Mrs. Wright Morris insists, in a communication to me, on depression-era prudence. The *Notebooks* offer a portion of Eiseley's otherwise unpublished wolf novel/beast fable, with a mad, deaf mother wolf (pp. 195–206).

marking of "their last earthly appearance" and *turning* of "turned away." Like the biological failures that inaugurated new lines of life, this turning-away that denied his mother and himself the secular immortality of descendants would seem to have yielded a kind of existential triumph.

Compare Said's incisive definition of the modern faces of the rhetorical problems of persona and ethos: "The two principal forces that have eroded the authority of the human subject in contemporary reflection are, on the one hand, the host of problems that arise in defining the subject's authenticity and, on the other, the development of disciplines like linguistics and ethnology that dramatize the subject's anomalous and unprivileged, even untenable, position in thought."[47] The correspondence of Eiseley's interiority at the time of writing, to his past interiority, and to his ongoing utterance, is authenticated by his sacrifice of life, by his pained testimony to that, and by the pain itself. One may compare, *inter alia,* the affirmation by Hemingway's Frederick Henry in the hospital that men may want some tricky way out of a bad war but they do not kick themselves in the genitals to achieve it. As to the subject's anomalous and unprivileged exteriority to thought, Eiseley dealt with that dilemma by embracing it: "Disciplines like linguistics and ethnology" are defined as many fairy rings; words in print or stone, and the very matrix of language, are affirmed to be as transient as cities. And he foresaw wild oak bursting "through the asphalt of [Philadelphia's] Market Street" (*All,* 149). But, as with Tate's sign, the fairy rings can harbor community.

There are early hints of the sense of sacrifice, as in a letter to his literary mentor, Lowry Wimberly, of the *Prairie Schooner:*

> My wife's friends are constantly being startled by the skulls which glare down at them from the top of one of my book cases! (We're short of room here in the apartment, and we've lived with dead men so long that we occasionally forget that outsiders must occasionally regard us with the horror and disgust which is reserved for New Guinea head hunters. . . . Anyhow the skulls don't bother us. I like to think they are really friendly and cozy up there on their shelf. Most of them wouldn't be there if I hadn't rescued them from situations that were taking them out of time altogether by the simple process of erosion. But I rescued them, treated their crumbling bones with alum and now they're safe and and [*sic*] warm till I, too go my way into the

47. Said, *Beginnings,* 293.

darkness. After that they'll have to look after themselves again. But for a little while they're safe. I like to think they know it, too.

Well, enough of playing Dracula.[48]

The letter is dated September 1 and must be from either 1943 or 1944. Eiseley's books were yet to be written, largely yet to be thought of. The skulls were quasi children, solicitously brought out of the dark and cared for, but necessarily left to their own devices. "Dracula," before scattering the seeds of his "concealed essays," drew from the skulls the blood of a poignant symbolic status.

Although Eiseley was ambivalent about taxonomy, resisted being simplistically schematized himself, and said so in letters and in print, it is possible to descry something of a rough progression from the uneasily playful semiconsciousness suggested by the letter, to the awareness of both general significance and acute particular anguish in his last writings.

In his three-part essay "The Inner Galaxy," Part 2 tells of his falling on a California street, presumably during his year in Palo Alto (1961–1962), and half-dazedly spontaneously addressing his own "blood cells . . . through my folly and lack of care . . . dying like beached fish on the hot pavement." Out of a "sensation of love on a cosmic scale," he said, "Oh, don't go. I'm sorry, I've done for you." Just afterward, in Part 3, he speaks of sitting on the beach beside the ocean—elsewhere regularly identified as the cradle of life—and feeling for creatures of air, land, and sea a "love without issue" (*Unex.*, 177–78, 191). Surely his sense of responsibility and conflict of feelings are together striking. The forces impelling him to spill his seed fruitlessly were almost as relentless and encompassing as the gravity bringing him down on that curb.

Writing not much later in an essay with the thanatopic title "Man in the Autumn Light," he tells of a kind of dark night of the soul in a place of the negative way: "My own race had no place in these mountains and would never have." Subsequent recollections of the scene moved him to reflect that "it was as though I carried the scar of some unusual psychic encounter." A physician friend's discourse of the body's reparative efforts is invoked for ambiguously metaphoric, metonymic purposes,

---

48. Loren Eiseley to Lowry Wimberly, September 1, 1943 or 1944, in Archives, Love Library, University of Nebraska. Quoted by permission of the Estate of Loren Eiseley.

because "mine was the wound of a finite creature seeking to establish its own reality against eternity" (*Inv.*, 121–23).

Near the end of *All the Strange Hours,* he has a figure he calls the Player say to him in a dream or reverie just after another vivid dream, "You played against all your possible futures. . . . You lost the unborn, remember?" He answers, "I remember" (p. 261). Two poems in *Another Kind of Autumn* venture a double effort of symbolic repair: the aestheticizing of verse and a displacement. The poem "Mars" suggests the ambiguity of spatial distancing as it tells how Eiseley found reminders in meditating ostensibly on NASA's Martian probe vehicle:

> tears . . .
> . . . fall for those unborn, unused, reactivated
> by the unrolling film that prints
> light where no eye exists.
>
> (*Ano.,* 49)

In "Two Hours from Now," he seems to pause on the ambiguity and ambivalence of both spatial and temporal displacement:

> . . . We were all lonely.
> I had no brothers. I have no children. Why do I write
>   to myself as
>   dawn is breaking
>   two thousand miles away?
> Nothing will be solved.
>
> (*Ano.,* 42)

Dawn is breaking in the East Coast time zone, where his wife awaits the return of her wandering Odysseus. And the skulls too await.

Finally, in *All the Night Wings,* a volume of early poems he arranged, and of some new poems he wrote, as he knew he was dying, the title poem tells of entering a cemetery with a girl at night long ago. It was on a dare; they were young. They found a gravestone inscribed, "Our baby":

> I think she wept. I followed her away.
> No name was there.
>
>   .   .   .   .
> I do not know where all my days have gone.
>
>   .   .   .   .
> . . . owls that query in night voices, why?

I cannot answer but I know who wept.
Still, still within the dark I hear that cry.

(*Wings*, 87)

Although more material could be marshaled, the implication seems sufficiently clear. Eiseley testifies again to the circumstances that bounded and delineated him, along with the Shepard-Eiseley line, as mankind in general may be bound. Yet he argues diversely and insistently that baneful mythologies reigning at large and the exigencies of economic life at home and the wounds of individual history within the self need not totally determine anyone. But even the effort of revealing the self in symbolic begetting may be risky and disquieting, as he wrote to his friend the physician Dale Coman: "After getting a book out there seems to come a period, at least to me, of revulsion. One feels exposed, naked, and I always have an impulse to disappear into the hills and hear no more about it." [49] He elsewhere accepts that the prophetic subverter of received boundaries and forms may occasion suspicion: "Like some few persons in the days of the final urban concentrations, I am an anachronism, a child of the dying light. By those destined to create the future, my voice may not, perhaps, be trusted. . . . I speak from . . . the original dispersals, not from the indrawings of men" (*Inv.*, 123). Yet like the bird or early man of his own reflections on himself, or like the Heideggerian man of a necessary complementary perspective, he does repeatedly come out of the reeds into the clearing of his essential self and his community. His repeated cry is both of and to sameness across the boundaries of form and difference. It disparages by implication all human pretensions to taxonomic totality or fixity. It testifies to something like loving transcendences of seeming limitations in nature, in the individual transcendences of the fugitive who does not remain simply that nor silently alone.

He speaks for Americans, immigrants all—even those who came across the land bridge from Siberia—who have no extended history here and who still, if he but move us to realize it, may have no issue. We may be blessed with children and grandchildren as he was not, but his genealogical line—scarcely shorter than ours, perhaps, in geologic time—and his writerly life prompt us to trust more to our mutual cherishing than to our erstwhile provision for the future.

49. Loren Eiseley to Dale Coman, September 19, 1973, in Dale Coman File, University of Pennsylvania Archives.

7

DEATH AND TIME, TIME AND
DEATH: EISELEY'S PROFESSIONS

There are furies in me . . .
furies of time assuaged
only with time.

—Loren Eiseley, *All the Night Wings*

A small boy lives in a house he feels to be socially isolated, a house his beloved father frequently leaves on business trips, a house he feels to be tomblike in the silence of his lone, deaf mother or disintegratively troubled by her rasping voice scolding her son or, worse, quarreling raucously with her husband. The small boy, as one of his earliest projects, makes small wooden crosses and paints them gold—markers for a private graveyard for small animals. Does he pity and commemorate those creatures as abandoned or scourged by life?

The boy, growing older, becomes aware that a certain visitor is an older half brother, that the visitor's mother was their father's first wife, who had been pretty and greatly loved and had died young. Does he more or less guiltily think about the lost, good mother? He becomes aware from veiled references and whispers of the "mad Shepards" in his mother's family line—a disintegrative notion not improbably bearing symbolic equivalence with death.[1] The boy becomes a young man but one whose maturity at that time he would later be careful not to exaggerate. He has various brushes with death at work and in evasive travel. And he endures the long contemplation of a death by tuberculosis that

---

1. For the idea of death equivalence, and of separation, disintegration, and stasis as categories of such equivalence, I draw on Robert Jay Lifton's masterly *The Broken Connection* (New York, 1979). Manuscript notes in the University of Pennsylvania Archives reveal that Eiseley considered calling his autobiography "The Other Player: A Chronicle of Solitude" before settling on *All the Strange Hours*. In 1956, he considered "The Reaching Out" before settling on *The Immense Journey*.

he overhears a callous doctor predict.[2] He experiences the intrusively abrupt and shocking death of a schoolmate. He watches the slow, appalling death of his father. Though not to death, he nevertheless loses, apparently, one greatly loved woman.

There is no need to assign relative impacts; it is impossible anyway. What is manifest to all serious readers in the published writings of the fully mature Eiseley, it seems, is a deep preoccupation with death and with its symbolic equivalents of separation, disintegration, and what is less important for him, stasis. His letters and his conversation are similarly marked by the preoccupation. It is not clear how consciously he connected his "profession of time" or his profession of writing with the threat of death and its symbolic equivalents. That he made a life-affirming profession of the study of time, in the context of death awareness so potentially disabling as his, is something of a triumph.

From the "immense journey" of vertebrate life, written in his forties, to the "strange hours" excavated from his own life, written in his sixties, he seems to have contemplated times and landscapes, including a nighttime museumscape of fossil crabs, with an auspicious eye or a shudder, depending on whether they could be integrated with some element of human being or not. A remark of Robert Jay Lifton's that at first encounter may seem too sweeping is in fact powerfully descriptive of Eiseley's professional engagement with time: "We learn the principle of time from interruptions in our unquestioned relationship to it . . . from early death equivalents."[3] Moreover, the connections he sought, the presences he tried to foster, should be understood in this context to mark not only actual losses and feared losses but also vivid imaginings of the never-possessed, counterweights to loss.

The writing career to which the present study attends figures forth during the last thirty years of Eiseley's forty-some years as a professional of time. There is from the first a retrospective cast to it, running from dissertation work (not otherwise noticed here) on the utility of retrospective instruments, that is, on *indexes* of time, to his very late decision to re-collect and republish poetry he had composed before the years of his technical paleontological writing. The reason could be stated in

2. As E. Fred Carlisle notes, two uncles died and "Uncle Buck" made a "well-nigh miraculous recovery" from tuberculosis (*Loren Eiseley: The Development of a Writer* [Urbana, Ill., 1983], 24).

3. Lifton, *The Broken Connection*, 138.

words that sound very like his own but that come from George Kubler's *The Shape of Time*, two copies of which he owned: "Men cannot fully sense any event until after it has happened, until it is history, until it is the dust and ash of that cosmic storm which we call the present, and which perpetually rages throughout creation."[4] Time is both something that happens to us and a symbolic model, like "nature"; like nature, too, it is partly a reflexive model of the formulating psyche and its anxieties and ideals. "What time is it?" seems always to have been a more resonant question for Eiseley than "Where am I?" since the question for him always reached beyond the mere hour of night or decade. He wrote early of the "enormous mindlessness of space," apparently more with a sense of the irrelevance of the vastness than of fear like Pascal's at its silence (*Imm.*, 25). He wrote late in life precisely to ascribe temporal quasi sentience to a remote datum, a pulsar identified by the Jodrell Bank radio telescope in 1953:

> If immortality
> is to outlast two universes collapsing inward,
>   and their renewal,
> then this creature in the constellation Cygnus
> is the terrible eye of all the past . . .

(Ano., 39)

The spatiality is important to him because it can lead to the notion of perspective and the concept of relative importance. As Kubler put it, "The historian and the astronomer both transpose, reduce, compose, and color a facsimile which describes the shape of time."[5] But for Eiseley it is the *changing* shape of time, that dynamic of potentially better or worse life, that is primary. Evolution continues and accelerates as culture. He applauds the two-hundred-inch eye of Mount Palomar Observatory and the eye of the electron microscope for balancing each other and helping to situate man conceptually "where he belongs; and that is amid differing orders of being and differing orders of temporal dynamism." But yet, for all such "beautiful machines," men may "have no true time sense, no tolerance, no genuine awareness of their own history. By contrast, the balanced eye, the rare true eye of understanding,

---

4. George Kubler, *The Shape of Time: Remarks on the History of Things* (New Haven, 1962), 18. Compare Eiseley's poem "Snowstorm," in *Inno.*, 111–12.

5. Kubler, *The Shape of Time*, 19.

can explore the gulfs of history in a night or sense with uncanny accuracy the subtle moment when a civilization in all its panoply of power turns deathward" (*Inv.*, 88). He would be *in mediis rebus* not only as *metaxy* but as hermeneuticist.

Aspiring to the true eye of understanding, moved by cultural occasion, impelled by biological inevitability, and sensitized by personal bent, Eiseley took time fundamentally as the medium of, or else as a metonymic figure of, death and deathly loss: a wound (*Night*, 221), a ravager of faces whether topographic or human (*Imm.*, 3; *Unex.*, 104), a burden (*All*, 5), a sinister attenuator (*Unex.*, 15), a dark chasm, mother, risky depth (*Night*, 15, 27, 157). Initially, it seems, and intermittently, time's baneful power might appear to permit not evasion but a sort of identification with the aggressor[6]—by some alignment with the consequences, as in the stasis of "The Most Perfect Day in the World," about the day "time stood still" (*All*, Chapter 7) or of the comment "I went [as a graduate student] to the gates of a deserted cemetery. With the sure instinct that time would vanish here" (*All*, 83), or by imaginative alignment with some state anterior to human time, as, *inter alia*, in his poem "The Black Snake": "Time stopped there and remained Eden" (*Ano.*, 55).

One can almost construe Eiseley's project, or less tendentiously, his coproject with self, as the Heideggerian interpretation of time as the possible horizon for any understanding whatsoever of Being.[7] However fundamentally time and Being may have been for Eiseley a horizon of death, separation, and disintegration, his signal response was to make of his double profession of time and writing something manifoldly other. Eiseley himself introduced *The Innocent Assassins,* the volume of poems he published in 1973, by saying, "As a young man engaged in

6. In something of that spirit, Andrew Angyal quotes an unused remark prefatory to *The Night Country:* "I must always have distrusted time and therefore I came to study it" (*Loren Eiseley* [Boston, 1983], 99). But the remark was unused by Eiseley; I think the issue is more complex.

7. Martin Heidegger, *Being and Time*, trans. John Macquarrie and Edward Robinson (New York, 1962), 19.

[paleontological work in the Wildcat Hills and Dakota Badlands], my mind was imprinted by the visible evidence of time and change of enormous magnitude. To me, time was never a textbook abstraction." He was affected by the vividly concrete textbook of nature, then, or more exactly, a press form imprinting his mind, making of *it* a sort of textbook. That would make the essayistic writings, including some emphatically essayistic poems, something like offprints. But this is too simple a model—a point worth returning to.

What is there of time as a nonabstraction in the nontextbook of Eiseley's writings? Initially striking to many readers is the rendition of a *dynamism* in time at odds with commonplace perceptions of it as glacially slow. A less striking and more slowly cumulative realization is that both the literal places of evident vastness and the symbolic places—the places set aside for studying it—could afford Eiseley needed respite and intervals of something like tranquillity. That this is crucial, Ortega reminds us in a remark Eiseley may have known but temperamentally would have conceived for himself: "Without a cetain margin of tranquillity, truth succumbs."[8] That this seems to relate paradoxically to the dynamism in time is something to which it will be necessary to return. A closely related point, not without its own paradoxical features, is that Eiseley's writings show the study of vast or at least remote time yielding something of the intimacies of home and family, intimacies so deficient or defective in his own early life as to leave a lifelong hunger. Yet it is also evident to careful reading, and predictable insofar as anxieties over death and death equivalents animated him, that paleontological time could provoke his darkest sense of separation and estrangement. But equally clearly, and contrariwise, time's long, *unreturning* "historical" record and the life-forms developing mysteriously in it could occasionally suggest some transcendence that Eiseley might freely express as freedom or constrainedly intimate as almost miracle.

The "*Space* that measures Day and Night" has ordinarily figured time in Western literature before Milton and since. But as the careful words about "time and change of enormous magnitude" suggest, and as Eiseley's whole book *The Immense Journey* and numberless subsequent remarks can confirm, the timely space of the paleontologist is energetically concrete and literal even while serving as an imaginative medium.

8. José Ortega y Gasset, *The Dehumanization of Art, and Other Writings on Art and Culture* (Princeton, 1948), 186.

At the beginning of *The Immense Journey*, Eiseley tells of kinesthetically working his way down a slit in stratified rock: "I was deep, deep below the time of man in a remote age" (*Imm.*, 4).

That relative mobility in—*in*—sensory surrogates for time exemplifies in one way his sense of dynamism as it corporealizes the effortful mobility of body and mind in natural history. Even more saliently, it abbreviates, as the figure continues through the essay, all mammalian and primate development. Scarcely since Sir Thomas Browne, whom Eiseley admired, wrote of Egyptian antiquities decaying "like pillars of snow" has anyone been so given to the stunning compression of the vastly extended in time and to the acceleration of the imperceptibly slow into images so readily dramatic in human awareness.[9] The passage and the whole essay in which it appears also exemplify control. Eiseley *climbs*—not falls—down; mammals and primates, as it were, *chose* forks in the road. The slit, which stands "symbolically . . . for . . . time" (*Imm.*, 11), can even be kiddingly made to represent death, but under gamelike control: "The Slit was a little sinister—like an open grave, assuming the dead were enabled to take one last look—for over me the sky seemed already as far off as some future century I would never see" (*Imm.*, 4). But he can and does climb out, and he deliberately attempts to fix the symbolization in writing.

Yet in his symbolization of accelerated temporal-spatial processes, he underscores a certain kind of letting-go. For example, he writes of floating one summer day in Nebraska's shallow, sandy Platte River: "Moving with me, leaving its taste upon my mouth and spouting under me in dancing springs of sand, was the immense body of the continent itself, flowing like the river was flowing, grain by grain, mountain by mountain, down to the sea" (*Imm.*, 19; compare James Hutton by Scottish brook, in *Dar.*, 65). This might have led to a cadenza on what he elsewhere calls the "mind as nature" or on his respectul awareness of regression in approximately Freudian terms ("that mother element which . . . shelters"; *Imm.*, 20).[10] But it does not, and need not for present purposes either. Again, he writes of "that far-off Cretaceous explosion of a hundred million years ago," the "soundless, violent explo-

9. For a similar view, with reference to "The Crevice and the Eye" (*All the Strange Hours*, Chapter 10), see Angyal, *Loren Eiseley*, 24. Angyal helpfully notes that "The Crevice and the Eye" derives from a piece in the *Nebraska Alumnus* of October, 1937.

10. Carlisle speaks of regression "clearly in service of the ego" in the 1931 poem "Earthward" (*Loren Eiseley*, 63).

sion . . . of the . . . flowering plants . . . like hot corn in a popper" (*Imm.*, 63, 72). What the left hand takes away in *far-off* and *hundred million*, the right hand gives back in the familiarly experienced abrupt blooming of flowers, and corn popping, and the implicit contrast with explosions not so ironically benign as this one. Like the lively "spouting" and "dancing" of the continent, this is countered by—Tate might have said, "is in tension with"—Eiseley's almost Renaissance vision of entropic mutability.

Yet more emphatically, he recalls, "I . . . had seen [creaturely] shapes waver and blow like smoke through the corridors of time" (*Fir.*, 176), and he writes of "that eternal flickering of forms" (*Imm.*, 138) "with little more consistency than clouds" (*Imm.*, 6). He remarks, "This lengthened time-span was peopled with wraiths and changing cloud forms. . . . Life . . . altered its masks [and it] can be seen . . . as a complete phantom" (*Inv.*, 14–15). The genetics of sexual reproduction, he says, give nature the "roiling unrest of a tornado. It is not the . . . palace of the Eighteenth-Century philosophers" (*Unex.*, 46). Leslie Gerber and Margaret McFadden suggest that some moments such as those—and suggest more convincingly that some of Eiseley's fictive narratives, notably "The Dance of the Frogs" (in *Star*) and "The Creature from the Marsh" (in *Night*)—are to be understood as "ventures in surrealism."[11]

Examples of temporal dynamism could be multiplied, their range perhaps somewhat extended. What is happening in such writing? An obvious point: "No one likes to watch, listlessly, an hour hand go around the clock." But Eiseley goes on with words of condescension, in a context of disparagement, toward nineteenth-century catastrophism: "We want the cuckoo bird to erupt violently at intervals from his little box, or a gong to strike. This catastrophism provided" (*Fir.*, 24). Yet, a page later, skirting his own contradiction and before gently discounting Hutton's world model as too like a heat engine, he praises the pioneering insight of Hutton in the eighteenth century: "For him and him alone, the water dripping from the cottagers' eaves had become Niagaras falling through unplumbed millennia. 'Nature,' he wrote simply, 'lives in motion'" (*Fir.*, 25). For the anticatastrophist Eiseley, whose uniformitarianism often takes the image of an immense journey on a long road with many turns and branches, the dynamism must seem paradoxical:

11. Leslie E. Gerber and Margaret McFadden, *Loren Eiseley* (New York, 1983), 128–38.

"Man and his rise now appear short in time—explosively short. . . . The profound shock of the leap from animal to human status is echoing still in the depths of our subconscious minds" (*Imm.*, 91–92).

Evidently, when Eiseley accelerates and intensifies the imperceptible processes of what we casually call nature, he is responding to human boredom that may trouble him as a threatening feeling of stasis at the same time that it can represent a disintegrative shock that he conceives to be generally human. Perhaps both sides of boredom are one, as Lifton seems to argue.[12] On the other hand, we find this permutation of the metaphor of journey: "If we were able to follow evey step of man's history backward into time, we should see him divested, rag by rag and stitch by stitch, of every vestige of his human garment" (*Imm.*, 105). Paleontology does enable something like a radar plot of the trajectory of time's arrow—or perhaps, thanks to the peculiarly Eiseleyan comradeliness of tone and vitality, something more like the teasing semipresence of a locale given through a varitinted 1 : 25,000-scale ordnance map.

Rare for Eiseley in published work, he twice uses the same extended locution in one book:[13] "Cat and man and weasel must leap into a single shape" (*Imm.*, 5, 160). He situates "those converging roads" (p. 5) as "inconceivably remote from us now, far back along the time stream" (p. 160). Somewhat like Byron's heroine, who without seeming to consent, consented, he, denying conception, conceived, in a characteristically mature effort to master experience never altogether masterable. One strategy was for him, at least formally, not any longer to be a participant but to act like the football coach or announcer who by expertise and video-replay technology explicates how a football play had one option forestalled, then a second, but finally succeeded by clever and daring recourse to a marginal third option. A bit later, in "The Time of Man," the forms "shrink to single tree shrew." Man "was and was not," so imponderable the possibility (*Mr.X*, 224).

One of his options, if not precisely a strategy, was that regressiveness which he might venture for as long as an afternoon on the Platte River or as a page in the telling but which he rejected as a way of life: "It is inevitable that transitory man . . . should entertain nostalgic yearnings

---

12. Lifton, *The Broken Connection*, esp. Chapter I.

13. But Eiseley very commonly, and economically, wrote in letters like a bard repeating or slightly varying oral formulas, as when he responded to what might be called generic fan letters, and occasionally in contemporaneous letters to different friends.

for some island outside of time, some Avalon untouched by human loss. Even the scholar has not been averse to searching for the living past on islands or precipice-guarded plateaus" (*Imm.*, 31–32). An alternative strategy is to be a player.

Play itself can be a viable response to the yearning, as in his verbal play about the gravelike slit, or in his bodily play with a fox cub, a "way to run the arrow backward" (*Unex.*, 210–11), or in the compound playfulness of a late poem: "I was regressing rapidly to / an American street in a small town where / I could not be followed / . . . Oxford, I think, is a place at the end of the world" (*Wings*, 86). The hostile deprivations and disintegrations and longueurs of life are dulcified by geological materialization and paleontological interpretation in an immense uniformitarian narrative. Turbulence and conflict can then be reinscribed in the textual representation of the book of nature as, notably, artistic foreshortening, which is to say, conflict under control (the degrees vary), somewhat as in a game. The book of nature interpreted moves with the interpreter's mind to Palo Alto or the "great cities of the East" or anonymous hotel rooms, where the reinscription may proceed.

One may distinguish from play, and also from effortful contol as from regression, the relief Eiseley sometimes found in extended time or figural space. Call it respite. It is respite from the "foreboding [that] still troubles the hearts of those who walk out of a crowded room and stare with relief into the abyss of space so long as there is a star to be seen twinkling across those miles of emptiness." Foreboding and relief—a quintessentially Eiseleyan odd couple. He speculates that Boskop men "may have lacked something of the elemental savagery of their competitors" (*Imm.*, 125, 138). He reflects in a poem that "the rock will close upon us / . . . deaths but not pettiness, / because not human. / . . . Extinction is an art too great for man, he bungles it / by obscene malice" (*Inno.*, 55–57). It is worth noting that he does not write, and it is a fair guess that he did not feel, the malice to be *inexplicable*. But it is more to the purpose to note his term for the process of natural extinctions: *art*. It suggests at least quasi purposiveness in the mysteriousness and unexpectedness of the universe. Its otherness may be embraced with a kind of trust as long as the trust conceals no condescending view of nature as mechanistic. He may have known directly, or from *Man and People*, Nietzsche's aphorism that Ortega's translator renders as, "We feel so tranquil and at ease in pure [*freie*] Nature because Nature entertains no opinion about us."[14] He writes in "The Intellectual Antecedents of *The Descent of Man*" that "no man afflicted with a weak stomach and insomnia has any business investigating his own kind. At least . . . until they have undergone the petrification incident to becoming part of a geological stratum" (*Mr.X,* 202).

Eiseley wrote from an excavation site, at twenty-five, to a college literary friend. Less guarded than most of his later letters even in the sweatiness of its artfulness and the striking of attitudes, it reveals a good deal about his restless ambivalences and even anticipates the metaphor of slime-mold that he developed thirty-eight years later in *The Invisible Pyramid.*

> Redington, Nebraska; 20 May, 1932
>
> Here where there are only bones asleep in their million-year long rotting, where the events of the Pliocene seem more real than those of civilization— that overnight fungus, you may take it as a compliment that I was genuinely moved by these sonnets. Love is a faint far cry along the wind here—an old troubling lament from the world's edge that came to me in your letter. Not real. Something suffered a long time ago before I stepped out of time. That is the way it is here.
>
> Or was. Yesterday a pretty young gypsy waved at me from a camp by the road. It took me a whole day to forget—not very successfully. Why? I don't know except that she was young and beautiful—and I was young—and we would never see each other again and never speak. And she was one of the outcast people among whom I should have been born—the people who have no ties but a duty to horizons, who never grow old as we do squatting by the little fire of memory that goes out and leaves us to freeze alone in the end. . . . Tell me again the name of that book in which a man escaped his world by way of a pedlar's cart. I must read something of the sort before I find myself taking root and turning into a yucca plant on one of these bare hillsides.[15]

He was distressed, and he stepped out of time into a landscape at once therapeutically empty and subject to intrusions, to brief images of

---

14. José Ortega y Gasset, *Man and People,* trans. Willard Trask (New York, 1957), 90; Friedrich Nietzsche, *Human, All Too Human,* Aphorism 508. I am indebted to my colleague Adrian del Caro for Ortega's source.

15. Loren Eiseley to Wilbur Gaffney, May 20, 1932, in Archives, Love Library, University of Nebraska. Quoted by permission of the Estate of Loren Eiseley. "That book" is (as Angyal notes) *The Golden Journey of Mr. Paradyne,* by William John Locke, author of *The Beloved Vagabond* and other novels available through the Lincoln City Libraries.

other selves troublingly beckoning back to life. Decades later, he could imagine a somewhat more acculturated remoteness, praising Bacon for leading men "not in the oceans of this world but in the vaster seas of time . . . toward a green isle . . . a new-found land greater . . . in time" (*Man*, 49, 52). By the end of his life he would write to another friend, after a brief field trip, "Somehow I am always ending up in the wilds of Nebraska, Montana, or other places which frankly I liked better when I was young."[16] But by then he had exhausted the paleontological landscape's resources for respite, except for the ironic sense he had indicated wryly enough when writing publicly half a dozen years earlier: "There is a paradox to all digging that only an archaeologist would understand. The best way to be resurrected is to be forgotten. Consider the case of Tutankhamen" (*Inv.*, 102). The essays and poems would constitute his golden mask, but never a bland one. Respite in paleontological time can be a "historic, novel, and unreturning" alternative (*Unex.*, 36) to the "new scientific determinism" (*Fir.*, 141), but it is well to recall his furies and that they might be fully "assuaged only with time," in the sense of death. "I have sought refuge in the depersonalized bones of past eras (*All*, 145).

And yet what reads like a fleeting and flawed respite *there* from the meanness or fret of personal experience unmistakably could open *here* into a greater thing: the possibility of apprehending extended time as the medium of extended family—extraordinarily extended and generously familial. More immediately, Eiseley belonged to "that elderly professor [Frank Speck] in somewhat the same way that he, in turn, had become the wood child of a hidden [Algonkian] forest mother" (*Unex.*, 63). With symbolic expansion, he in 1956 interpreted an evidently ceremonious Neanderthal burial that included artifacts: "Across fifty thousand years nothing has changed or altered in that act. It is the human gesture by which we know a man, though he looks out upon us under a brow reminiscent of the ape" (*Mr.X*, 199–200). He quoted the remark three years later, to revise it delightedly in the context of a similar discovery by Louis Leakey: "An aspect of that act has been made distant from us by almost a million years . . . a humanity—if this story is true—that runs well nigh as deep as time itself" (*Fir.*, 113–14; on healed compound facial fracture in skull from "ice-age gravels," see

16. Loren Eiseley to Patrick Young, July 31, 1975, in University of Pennsylvania Archives.

Night, 157). Time, here as elsewhere, is insistently an unnatural human idea, like nature; whatever happened before protohuman consciousness, although it left evidences we can sometimes (we believe) interpret very well, was of an outer, less familial circle. More generally and metonymically: "The species alter . . . but the *Form,* that greater animal which stretches across the millennia, survives. There is a curious comfort in the discovery. In some parts of the world, if one were to go out into the woods, one would find many versions of oneself. . . . It is almost as though somewhere outside, somewhere beyond the illusions, the several might be one" (*Fir.,* 82). More particularly and metaphorically, if withal playfully:

> On my office wall is a beautiful photograph of a slow loris with round, enormous eyes set in the spectral face of a night-haunter. . . . Only a specialist would see in that body the far-off simulacrum of our own. Sometimes when I am very tired I can think myself into the picture until I am wrapped securely in a warm coat with a fine black stripe down my spine. . . .
>
> At such times a great peace settles on me, and with the office door closed, I can sleep as lemurs sleep tonight, huddled high in the great trees of two continents. Let the storms blow through the streets of cities; the root is safe, the many-faced animal of which we are one flashing and evanescent facet will not pass with us. (*Fir.,* 85–86)

Eiseley's sympathetic identification finds kinship with what is taxonomically remoter, too: "The archencephalon / . . . the nose-brain of our reptile past, / . . . Let him breathe / autumn; / he is part of myself" (*Inno.,* 23). He comments that the move ashore was made by the "failures" of the sea, and that the old-fashioned tree-dweller "was much too late" for either teeth or hooves—"a ne'er-do-well, an in-betweener" (*Imm.,* 54, 74–75). Partly the intellection here is steered, obviously, by fellow feeling, for Eiseley's later writings, notably "Obituary of a Bone Hunter," among others in *The Night Country,* and *All the Strange Hours,* make clear that he tended intermittently to think of himself as some sort of a failure. The dynamism of Eiseleyan temporal foreshortening can give the in-betweeners, the clamberers into genetic niches, and the creepers through rents in the genetic curtain something of the hectic air of television escape artists.

It could not be otherwise. Nature is marked by—Eiseley reiterates the conventional phrase—great extinctions. Their mechanism may be rather more probably explicable now, although it is still apparently subject to argument. A somewhat better understanding exists of the mecha-

nism of killing winters from dust extraordinarily cast into the atmo-
sphere, and of solar flares (on which he speculates in *Invisible Pyramid*)
and the ozone layer. But he is more concerned with the situation of any
such knowledge in an ethical and sociopolitical continuum: "In the first
four million years of man's existence, or, even more pointedly, in the
scant second's tick during which he has inhabited cities and devoted
himself to an advanced technology, is it not premature to pronounce either upon his intentions or his destiny?" (*Inv.*, 93).

Still, the fellow feeling could rasp or fail. He has a late poem that
meditates on the finding of the Dead Sea Scrolls and on part of one
scroll's message in a way seemingly owing something to the literature
and practices of meditation of the seventeenth century. He quotes the
scroll: "Thou has made a mere man to share knowledge / a thing
pinched out of clay." His poem continues, "To share, to be companioned? / How shall I be companioned?" (*Inno.*, 97). It may be the most
urgent question of Eiseley's adult life, to which the dedications to his
books offer several more or less expectable answers. Their contents give
other similarly particular answers. But almost every essay answers, too,
that his companionships in the frames of personal life, species life, biological life over geological time—all stood in vulnerable contrast to
separation and estrangement. In the poem just cited, he predicts a world
desert and prescribes artful words only for the short-run millennia, the
kit fox for the long run. He finds loneliness in dissent from the guild of
dedicated scientists (*Imm.*, 206), finds man "physically antique in this
robot world he has created" (*Imm.*, 89), and finds that the "thread" of
life "all the way back into Cambrian Time" (*Fir.*, 56, 54) presents an
immense vista more uncompanionable than companionable. Eiseley
emerged from a cave early in his professional life to find—and to find
ever thereafter—the cultural world metamorphically alien (*All*, 105).

"The Last Neanderthal" presents in its second section an extended
and moving recognition of primate connectedness over time, evoking
the recognition so ambivalently in its twentieth-century setting as to invite, paradoxically, the reader's sense of unsettled disjunction. In the
narrative and meditation, Eiseley remembers a girl who delivered milk,
butter, and eggs to his fossil-excavating party on her father's hardscrab-
ble hill-country farm in western Nebraska. Eiseley gradually discerned,
or fancied, a genetic reappearance of *Homo sapiens neanderthalensis* in
her heavy body, large head, and archaic brow ridges, yet at the same
time, he sensed a strongly companionate human creature. "Retarded,

you'd say," remarked the late Emery Blue, a judicious scientist, a friend of Eiseley's, and a fellow member of Eiseley's party.[17] Perhaps I would. Yet how little we know of mental retardation, especially of that which does not follow prenatal disease or other insult to the fetus but simply appears. Simply? One observes that many designated by the holistic term *retarded* and evidencing substandard capability in some functions of mind may exhibit standard or better capability in other functions. If Eiseley's term *Neanderthal* needs to be understood as a metaphor, it is at least provocatively alternative to *retarded*. Insofar as his word is not a metaphor, insofar as the gene pool cast up on a culturally barren Nebraska butte an atavistic form—like one he remembered from some dissecting table in his own experience—the disentangling of cultural blight and creaturely companionability is tenable if unprovable science. Recent scientific speculation has supposed high linguistic and visuospatial competence in *neanderthalensis*.[18] In any case, the girl communicated to him and through his written reflections the "utter homelessness of man" (*Unex.*, 227).

Opposite William Hazlitt's generic young man, who never really believes that he is going to die, Eiseley seems to have felt constantly and even urgently from his twenties onward that we are all for the dark, the autumn, the ground, the ashes, the "eternal storm" (*Inno.*, 112). He could aver that "we hear nothing / but the echoes from a deserted universe" (*Inno.*, 21). So much for logocentrism? Yes and no; echoes presuppose originating sound, and we see him read companionship across the millennia in the burial observances of early man. He also read in the fossil record that "vast desolation and a kind of absence in nature invite the emergence of equally strange beings or spectacular natural events. An influx of power accompanies nature's every hesitation; each pause is succeeded by an uncanny resurrection" (*Unex.*, 96). His word *uncanny* is not casual. Compare: "Somehow in the *mysterium* behind genetics, the tiny pigmented eye and the rocket capsule of Pilobolus were evolved together" (*Inv.*, 76). He affirms, too, that the human mind "so involved with time, moving against the cutting edge of it . . . can transcend time,

17. Emery Blue, interview with author, Lincoln, Nebr., July 1, 1985.
18. See Ralph L. Holloway, "The Poor Brain of *Homo sapiens neanderthalensis:* See What You Please," in *Ancestors: The Hard Evidence*, ed. Eric Delson (New York, 1985), 319–24, esp. 320.

even though trapped. . . . We are not wholly given over to time" (*Fir.*, 165–66).

He could not say exactly what the possibilities are for the human mind, although he would write up the uncanniness he found in nature and note the aspiration he found in himself. Near the end, he could report, "Ironically, I who profess no religion find the whole of my life a religious pilgrimage" (*All*, 141). He crossed out "Man: The Time Polluter," "The Time Stealers," and "The Time Collectors" to entitle Chapter 24 of *All the Strange Hours* "The Time Traders." It seems he never despaired of our coming "to know ourselves, our past, our own true hunger, and where the real frontier we seek lies waiting."[19]

Lifton argues in a sentiment Eiseley seemed half-consciously to anticipate that "the attempt to exclude from the psychological imagination death and its symbolizations tends to freeze one in death terror, in a stance of numbing that can itself be a form of psychological death."[20] Nothing is more characteristic of Eiseley than that he tried to serve his country—as he tried to serve it in uniform (*All*, 142)—by proclaiming the time of all life's history to be environed with death, and the "frontier we seek" to be implicated with death, and the technology we may overtrust to be permeated with death. Somewhat as Ortega scolded Nietzsche for the superfluity of his injunction to live dangerously, Eiseley rebelled at the tendency of Lévi-Strauss's writings on his shelves to give "every problem, no matter how small . . . explicit delimitation."[21] He is more like Heinrich Böll's artist, who "carries death within him like a good priest his breviary." Death is the problem not itself delimited, and the mocker of many another delimitation. To taxonomize and detemporalize, as insufficiently careful structuralism might do, or to prejudge, as many of his professional colleagues would do, the question of what

19. Loren Eiseley, Introduction to *The Shape of Likelihood: Relevance and University,* ed. Taylor Littleton (University, Ala., 1971), 18.

20. Lifton, *The Broken Connection,* 8–9.

21. The phrase is Edward Said's, of structuralism in general (*Beginnings: Intention and Method* [Baltimore, 1975], 323).

Eiseley just once went so far as to call "the eternal mystery, the careful finger of God" (*Imm.*, 52) would be to succumb. To become idolatrously empiricist would be to succumb to the death of numbness, as he evidently concluded in the life crisis densely emblematized in his essay "The Star Thrower," which was discussed in the previous chapter. But what is the alternative? How is it possible to take imaginative control of that which exceeds other control or to take imaginative notice of that which by definition exceeds definition? What control is lively, rather than deathly overcontrol?[22]

The moment of innocent play, the gesture of loving trust (as of tenderness across the millennia in burial practices, or accord with a fox cub or "talking cat" across species), and observations of mysterious complexities in nature—all can invigorate the imagination and position the self for action: "Brains and sympathy—the mark of our humanity—will alone have to guide us. The precedent of the forest will be wrong, the precedent of our dark and violent mid-brains will be wrong; everything, in short, will be wrong but compassion—and we are still the twofold beast" (*Mr.X,* 228).

The other ground for imaginative control of the fears and despairs about death is the idea of chance, which may allow escape for at least a genetic line, which eventuates in the radical statistical improbability of each human individual, and which can subvert the determinisms of nature or culture: "There is a part of human destiny that is not fixed irrevocably but is subject to the flying shuttles of chance and will" (*Inv.,* 55). Eiseley could at the end of his life reflect more generally on the significance of the idea, albeit with mention of more particular chances: "I seem preoccupied with chance, whether it be the chance that determines life or death upon a street corner, or what it may have been that hovered about me in the ruined farmhouse where, as a child, I threw dice" (*All,* 248). Chance offers a parodic alternative to the argument from design, which he evidently finds rationalistic and reductively inadequate to a Haldanish universe "queerer than we can suppose." Design may imply in notoriously ambiguous fashion a Designer,[23] but chance may imply,

22. Gerber and McFadden speak of Eiseley's perceiving the "will to control" as robbing "scientists of their inner life" (*Loren Eiseley,* 100).

23. "The Coming of the Giant Wasps" (in *All,* Chapter 23) is clearly, among other things, a counterpart to Robert Frost's poem "Design," with its own "design of darkness to appall."

as he seems to suggest in representations of dream or reverie, a Player.[24]

But against the occasional triumphs of life as if on wings (*Imm.*, 167), against the benign plays of the Player and human improvements of the odds and faithful assertions that "there are aspects of the world and its inhabitants that are eternal" (*Inv.*, 55, 112), and against adversions to life-forms benefiting by a niche or rent in the ecological screen, there is the textual drumbeat, the iterative punctuation, of death. From no rent in any screen did any child of Eiseley issue forth to life, as he repeatedly notes. His ambiguous eternity seems apter to mock than to reassure. He represents himself as a young man and species typical (*Inv.*, 123). Species life may be mocked by glacial winter (for example, *Unex.*, 103; *Inv.*, 71) or by longer-lived species such as the hyena (*Inno.*, 93). The last example comes from a flawed and awkward late poem not lacking in feeling and a certain art. The five million years of the hyena's mocking presence as a species are implicitly mocked by the eternity of absence in death, which is likewise a mocker of human and individual efforts to connect in the uniquely human mode of the symbolic: "I . . . buried what I loved . . . in the red ochre that might bring her back. . . . Ochre is blood, it did not bring her back." Ochre "is" not only blood and hence associated with life, it is associated with reproduction.[25]

As he early put it, in an insight scientifically orthodox but expanded to prophetic discontinuity with his surrounding culture, the eighteenth- and early-nineteenth-century recognition of species death, phylogenetic death, was prerequisite to the linked recognitions of "drastic organic change" and ongoing creativity, without which imagination would be limited (*Fir.*, 33–34). Since imagination is intrinsic to our lives as persons, it can give us the constant sense of loss of others and of self, of the "blue worm," of the "seeds," of physiological and psychological death-wardness in the self (*All*, Chapter 21; *Star*, 210). When Eiseley wrote in 1960 that with Sir Charles Lyell's uniformitarianism, "death . . . was becoming natural—a product of the struggle for existence" (*Fir.*, 51), he meant by death, ostensibly, species extinction. But at a different level and in a different sense, he meant that we either die of unintegrated

24. Note related imagery of gates and doorways identified (especially in *Man*) as somewhat accidentally discerned escape hatches.

25. This is according to Bruno Bettelheim's *Symbolic Wounds: Puberty Rites and the Envious Male* (Glencoe, Ill, 1954), 168–69, 180. Eiseley owned that book. The lines are from "Deep in the Grotto," in *Inn.*, 109–10.

identity, absence, stasis, or die of being what we more largely are, painfully evolving aspirants. That is the "wound of time" (*Night,* 221). As Yeats would have an adult do, he conceived of life as tragic.[26]

26. This is a revision and expansion of a paper, "The Overdetermined Time of Loren Eiseley," read at the Modern Language Association convention, December, 1984. The view here, and my emphases, differ from but also, I believe, reinforce Carlisle's fine depiction of an Eiseley withdrawing into writing as an alternative to withdrawing from society—and even, I would say, from life (*Loren Eiseley,* esp. 15–16, 36–37, 186). Especially with regard to "the zero," see also Erleen Christensen's suggestive meditation "Loren Eiseley, Student of Time," *Prairie Schooner,* LXI (1987), 28–37.

8

THE CHRESMOLOGIC ESSAY

Philosophy springs from the love of being; it is man's loving endeavor to perceive the order of being and attune himself to it. Gnosis desires dominion over being; in order to seize control of being the gnostic constructs his system. The building of systems is a gnostic form of reasoning, not a philosophical one.

—Eric Voegelin, *Science, Politics, and Gnosticism*

The moment we cease to hunger to be otherwise, our soul is dead.

—Loren Eiseley, "The Time of Man"

What appears in an attempt to shift the figure-ground relationship, highlighting what has been obscured or occluded?[1] The scientist professor of time, of time furious or assuaging, professed in personal essays a "continuing biography."[2] In previous chapters, I have taken some notice of biographical matters, autobiographical moments, and postures of address. I have considered more prominently, though, the rewards and distresses to be found in human time by someone who early glanced approvingly at "Kierkegaard's faith in the eternal," as the "only way of achieving victory against the corrosive power of the human future" (*Fir.*, 121).

I have not highlighted, because I do not find, much reliance by Eiseley on the word *eternal;* the concept under that or other signifiers which I do find him insisting on is that of time not circumscribed by myth. His

1. For my own use of this familiar metaphor, I am drawing gratefully on reflections enabled by James Bunn's *The Dimensionality of Signs, Tools, and Models* (Bloomington, Ind., 1981).

2. Andrew Angyal, *Loren Eiseley* (Boston, 1983), 93. Angyal is quoting from a personal interview.

faith was less in the eternal than in geologic spans of time. I have tried to show his profound rejection of a *corrosive* human *future*, which Eiseley construed as a crushing commoditization of spatiotemporal perceptions, and as a taxonomizing in the time-space of science and culture more generally. He spoke of an opposition of technological time and "genuine earth time" (*Inv.*, 111), and I identify his quarrel as being with atomistic and static event time, or *tychastic* time, chance time.[3]

He writes in a very late poem, "The Eye Detached" (*Ano.*, 58), of the electronic viewer sent to the surface of Mars, extending quasi-human vision to a symbolic breaking point. It seems, so to speak, a vista where "nothing has died, been born, since yesterday's / four billion years," in contrast to the more intimate eye, turned hither, to "heartbreak Earth, so loved, so multitudinously / extinct." He certainly means that vision cannot be, dare not approximate too closely to, the merely mechanical, and that it is the "thread of life" and the human consciousness of that which gives time its duration. More generally, he was close to enunciating in that late poem the implicit rationale of the essays: that only love can redeem time's extent from empty and even mathematical abstraction to give positive meaning to geologic time and to eternity. Somewhat as for Milton in *Paradise Lost*, it is love alone that transcends time. But that point surely is not what first registers with readers of Eiseley's essays step by step.

His usual essay shows its nature quite markedly in connection with two matters: presence and linearity. Readers are implicitly invited to ask, What sort of presence *to* whom? Presence *of* what or *of* whom? Linearity of what? In his awareness that "the writer's audience is always a fiction," Eiseley is quite Ongian. In his dual sense—of that condition of fictive address, and of language's limitations—he should be thought of as incipiently Derridian or Foucauldian. In what he did generally as a writer, and in what he did most particularly in his scientific essays after 1947, he became something of an Ortegan—rather than Lévi-Straus-

---

3. See John Michael Krois, "Ernst Cassirer's Theory of Technology and Its Import for Social Philosophy," *Research in Philosophy and Technology*, V (1982), 209–22, esp. Part 4.

sian—*bricoleur*. He arranged and juxtaposed careful selections from what Ortega called *circumstance*. He is known to have had some acquaintance with the writings of Ortega, Lévi-Strauss, and Michel Foucault, whereas he is not known either to have read or not to have read Walter Ong or Jacques Derrida,[4] but my point would not be to assert direct or simple influence even if there were such a thing. Rather, the effort I make is, with some aid from Eiseley's other writings and with some attention to his change over time, to detect the nature and dynamic of the "concealed essay" he writes.

In *All the Strange Hours*, he tells of an ear infection that made him deaf for several months in 1948. The deafness coincided with a magazine's reneging on a commission for a "straitly defined scientific article." In the previous year, his essay "The Star Thrower," intimative of personal crisis and of the "renunciation of my scientific heritage," had appeared in *Oceans*, and his significantly entitled "The Obituary of a Bone Hunter," in *Harper's*.[5]

There seems scant reason to doubt that he could reconstruct over a quarter of a century later the train of thought that led him to turn the aborted article and its ilk into the essays he sequenced and published as *The Immense Journey:*

> Why not turn it—here I was thinking consciously at last about something I had done unconsciously before—into what I now term the concealed essay, in which personal anecdote was allowed gently to bring under observation thoughts of a more purely scientific nature?
>
> That the self and its minute adventures may be interesting every essayist from Montaigne to Emerson has intimated, but only if it is utterly, nakedly honest and does not pontificate. In a silence upon which nothing could im-

4. Eiseley lists Claude Lévi-Strauss's *The Savage Mind*, and José Ortega y Gasset's *Concord and Liberty* in the bibliography to *The Invisible Pyramid;* he owned Ortega's *Man and Crisis* and *The Origin of Philosophy*, Lévi-Strauss's *Structural Anthropology* and *Tristes Tropiques*, and Michel Foucault's *The Order of Things*. The unpublished (but photocopied) Inventory of the Papers of Loren Corey Eiseley in the University of Pennsylvania (admirably prepared by Caroline Werkeley and revised by F. J. Dallett) highlights what the papers in their profusion make obscure: he not only revised in the ways usual for any writer sensitive to language and audience but often and variously dismantled, quarried, and recycled his writings.

5. In the late lecture "Man Against the Universe," Eiseley spoke and wrote, or more exactly wrote and spoke and published, of better appreciating Emerson "when I had abandoned certain of the logical disciplines of my youth" (*Star*, 177).

pinge, I shifted away from the article as originally intended. A personal anecdote introduced it, personal material lay scattered through it, personal philosophy concluded it, and yet I had done no harm to the scientific data. (*All*,
177–78)[6]

Eiseley had a remarkable boundary experience here, in which a personal transition was intensified by his months of silence and textuality
except for his pleasure of talking interiorly to himself. We overhear, almost as interiorly.

The resultant sort of essay can scarcely be more concealed than
Eiseley's account of its etiology is ironic. Matters "straitly . . . purely
scientific" will at the very least be resituated and put "under observation." Will it become a voice in a chorus or an instrument in a concert or
a taxonomized exhibit? No, it will stand in quasi-spatial but dynamic
juxtaposition with the gigantic X-factor of personality, ironically and
reductively referred to by Eiseley as anecdote and material. What it
meant to be a person in post–World War II America was evidently an
increasingly preemptive question for him during the last thirty years of
his life, a time he saw as increasingly scientized and technologized. The
situation invited him to consciously tentative answers and to ironic reactions that may not always have been conscious.

One way for science to remain unharmed, thus situated and whatever
its purity, is for it to be understood as intrinsically no less problematical
than its surroundings. Eiseley both celebrated Darwin as an intellectual
hero and champion of the nineteenth century and uncovered an insufficiently acknowledged progenitor in the figure and writings of Edward
Blyth.[7] In *The Firmament of Time,* he wrote, "In [man's] brain there is
really only a sort of universal marsh, spotted at intervals by quaking
green islands repesenting the elusive stability of modern science—is-

6. See Angyal, *Loren Eiseley,* 38–40. In commenting there on this passage, and subsequently applying it, Angyal seems to me to underrate Eiseley's wiliness. Leslie E. Gerber
and Margaret McFadden explicate the essays in terms less of this passage than of six
"motifs" (*Loren Eiseley* [New York, 1983], 50–53). Carlisle honors the passage, and develops his anatomy of the essays in terms of four "layers or dimensions," which are "science, autobiography, figuration, and metaphysics" (*Loren Eiseley: The Development of a
Writer* [Urbana, Ill., 1983], and the preceding chapter). I see Carlisle's four layers or dimensions more as permutations and so remain uncomfortable with the degree to which
his language rigidifies and separates the allotropic and dynamic Eiseleyan argument.

7. What Eiseley wrote on these matters, in a sort of companion-sequel to *Darwin's
Century,* is helpfully collected in *Darwin and the Mysterious Mr. X.*

lands frequently gone as soon as glimpsed" (p. 177). He concluded a friendly letter to a science-fiction writer in substantively ambiguous and syntactically distanced fashion: "I might say that to me no science fiction can be more exciting than the relatively limited knowledge we have of our universe and the evolution of man and his earlier relatives."[8]

In days that were for him pre-Foucauldian, he wrote with an almost Hegelian awareness of the ambiguously constructive sense of the words he would be privileged and constrained to use. Words will sometimes fictively *still* the dynamic in lived experience or generalize away crucial particularities. He considered in one instance how it would have been "if there had been such a magical self-delineating and mind-freezing word," and he decided that "at any step of the [primate evolutionary] way, the word 'man,' in retrospect, could be said to have encompassed just such final limits. Each time the barrier has been surmounted. Man is not man" (*Night*, 54). On another occasion he wrote, "There is always the danger, as you know, of unwittingly hypostatizing 'man' or his 'brain' into a single object."[9] As with Tate, so with Eiseley, the sign is an ongoing transaction between a renewable community of understanders and a problematic sector of referentiality. It is far more complexly dynamic than even complex and reversible organic reactions, and it is certainly not to be confused with the clarity and fixity of print in general or Saussurean ovals in particular.

He wrote his essays in keeping with serious reflection on the implications of language as the crude but essential message bearer and connector. He knew that such implications included, besides the other incommodations of words, their costliness in presentational and representational loss and their vulnerability to idiosyncratic distortion. One is apt to pause in annoyance at some of his own uses, for example, of the word *far* as a hybrid sign of distance, estrangement, and projected loneliness or "blunted pain."[10] He comments extensively on the matter:

8. Loren Eiseley to Thomas F. Monteleone, December 19, 1974, in University of Pennsylvania Archives.

9. Loren Eiseley to Dr. Leonard R. Sillman, August 17, 1970, in University of Pennsylvania Archives. See also the word *dog* in "The Innocent Fox" (in *Unex.*, and *Star*), Section II. And see the suggestion with regard to both signifier and signified that "nature itself [is] a mindprint beyond our power to read or to interpret" ("Walden: Thoreau's Unfinished Business," in *Star*, 198).

10. The phrase is Julia Kristeva's for the style of Marguerite Duras, in "The Pain of Sorrow in the Modern World: The Works of Marguerite Duras," *PMLA*, CII (1987), 140.

"Objects and men are no longer completely within the world we call natural—they are subject to the transpositions which the brain can evoke or project. . . . In the attempt to understand his universe, man has to give away a part of himself which can never be regained—the certainty of the animal that what it senses is actually there in the shape the eye beholds. . . . Linguists have a word for the power of language: displacement . . . the ability to talk about what is absent" (*Inv.*, 142–45). For present purposes, I am taking consciousness to be linguistic, and the displacement double: of the sign user from the sign receiver, and of the sign from the referent. Eiseley notes in a similar context that "we exist in an inner solitude . . . a loneliness which is the price of all individual consciousness" (*Inv.*, 48).

But loneliness, the pain of disconnection, is "linked to the nature of life" as well, rather than being altogether disjunctive, and Eiseley was increasingly conscious of "writing so as not to die." [11] In a meditative essay that appeared in *Holiday* in 1962, he dealt with the Easter Island faces as cryptic texts, glossing them as "stones in which men tried to inscribe their immortality," as we sculptors or writers or builders ultimately shall have done (*Star*, 72). In letters he wrote during 1970 he refers to one of his books as his "brain child," to two of them as his "children"; later he writes of having "no children but these little marks on paper" and of not wanting them to "be jerked about too unkindly." [12]

He can invoke a text: "Man's first giant step for mankind was not through space. Instead it lay through time. Once more in the words of Glanvill, 'That men should speak after their tongues were ashes, or communicate with each other in differing Hemispheres, before the Invention of Letters could not but have been thought a fiction'" (*Inv.*, 63). On the other hand, though Joseph Glanvill was arguably as hyperliterate a textual man as Eiseley and a better Latinist, presumably, he wrote as someone heavily implicated in residual oral culture. Written words "speak"; writing was perhaps conceived by him, as by most of his com-

---

11. The latter phrase is Foucault's, from Blanchot, in "Language to Infinity," in *Language, Counter-Memory, Practice: Selected Essays and Interviews,* trans. and ed. Donald Bouchard and Sherry Simon (Ithaca, N.Y., 1977), 53. But I found it first in Fred G. See's "'Writing So as Not to Die': Edgar Rice Burroughs and the West Beyond the West," *Melus,* II (1984), 59–72.

12. Loren Eiseley to Mrs. Chase W. Love, February 17, 1970, Loren Eiseley to H. H. Howard, January 22, 1970, Loren Eiseley to Charles Plymell, April 25, 1975, all in University of Pennsylvania Archives.

patriots, as displaced or second-class speaking.[13] With Eiseley, in contrast, the emphasis is on writing, even though the sign may be a "nameless name upon the page" (*All,* 22). Son of a loved and sometimes oratorical but often absent father and a mother who was certainly deaf and who seemed to him crazily difficult, he came as a boy to feel that he "had to rely on silence" (*All,* 166). A frightened "toddler," he once stilled his parents' quarreling by "pleading wordlessly" (*All,* 24). He became the lifelong student of signs inscribed or fossilized in stone, to be silently scrutinized in essentially private acts of reading. He came to conceive of not only his but "man's story, in brief," as "essentially that of a creature who has abandoned instinct and replaced it with cultural tradition and the hard-won increments of contemplative thought" (*Unex.,* 7). The abstract category of tradition and the notion of contemplation by increments implied to him, as they do to us, writing and literacy and the kind of linear thinking sponsored by literacy: "Writing . . . was to open to man the great doorway of his past" ("The Long Loneliness," in *American Scholar* [1960]; also in *Star,* 16). His was indeed a silent, visual archaeology.

Of course, elements of tradition and increments of contemplation can be talked about or lectured about. Eiseley lectured about them; he preferred to do so, however histrionically, from carefully written texts. Just after his remark about man's acceptance of tradition and contemplative thought, he wrote, "Without writing, the tale of the past rapidly degenerated into *fumbling* myth and fable" (my emphasis). He cites as a decisive confirmation that "man's greatest epic, his four long battles with the advancing ice, . . . has vanished from human memory without a trace." [14]

13. On "residual orality" in the Renaissance and persistently in human life, see Walter Ong, *Ramus, Method, and the Decay of Dialogue* (Cambridge, Mass., 1958); Ong, *The Presence of the Word* (New Haven, 1967); Ong, *Rhetoric, Romance, and Technology* (Ithaca, N.Y., 1971); and Ong, *Orality and Literacy: The Technologizing of the Word* (New York, 1982). For the by now widely current arguments about "logocentrism," see, for convenient examples, Jacques Derrida, *Of Grammatology,* trans. Gayatri Spivack (Baltimore, 1976), and Derrida, *Speech and Phenomena, and Other Essays on Husserl's Theory of Signs,* trans. David B. Allison (Evanston, Ill., 1973), as well as the fine introductions to those translations, by Spivak and Newton Garver, respectively.

14. That is largely true, I suppose. Yet I wonder if Eiseley was consciously rejecting the notion of the Eden story as a sophisticated condensation and metaphorization of the last interstadial's giving way to the most recent ice advance (and to biologically increasing cranial size?). Ten years later he speculated on a solar flare and memories that last long when "clothed in myth" (*Inv.,* 27).

He treats the "fumblings" of voice in storytelling gently in his earlier work and scornfully later. He evidently agreed with Ortega that "every word as a spoken word is first of all an adverb of place": [15] " 'We're here, we're here, we're here.' . . . From the heights of a mountain, or a marsh at evening [man's voice] blends, not too badly, with all the other sleepy voices that, in croaks or chirrups, are saying the same thing" (*Imm.*, 25). But later he declined to interview for a biography: "Tape recorders are anathema to me. . . . [My work] may please you more if you are not forced to face tapeworn garrulities." [16] *Garrulities* suggests both the ultraliterate's impatience with the copiousness of orality and *some* deflection onto himself of his increasing impatience with—as he conceived it—the none too friendly who would intrude upon his "mind," that is to say, a "parti-colored illusion which hates to be pinned down." [17] He felt "less and less inclined to go out simply because I feel increasingly averse to that much human contact." [18]

On the other hand, though Eiseley may write in a cultural situation significantly farther from primary orality than Glanvill did, and may write as individually more appreciative of writing's double displacement from referents and from human contact, he cites him in such a way as to make of him a quasi-fraternal presence in his text, much as he cites Francis Bacon and Sir Thomas Browne, his seventeenth-century spiritual cousins, and sometimes Charles Darwin, Henry David Thoreau, and others. Some would wish to say that he makes of Glanvill a quasi presence, perhaps. But it was his way to overwrite the visual and perchance intrusively noisy physical presence of others, the "human contact" he so often shrank from, with the symbolically mediated presence that is part of the signified of all his signifiers. He could have said, "I write *far,* signifying alienation, but if you read it sympathetically, you have in a cer-

15. José Ortega y Gasset, *Man and People,* trans. Willard Trask (New York, 1957), 167.

16. Loren Eiseley to Alex Rode, September 26, 1974, in University of Pennsylvania Archives.

17. Loren Eiseley to William P. Albright, February 26, 1971, in University of Pennsylvania Archives. At least twice he rebuffed inquiries with the adage "Let there be some uncertainty about your departure": Loren Eiseley to Hugh Gilmore, September 1, 1970, Loren Eiseley to Robert Shekter, June 25, 1973, both in University of Pennsylvania Archives.

18. Loren Eiseley to Patrick Young, December 1, 1975, in University of Pennsylvania Archives.

tain measure symbolically joined me there." Correspondingly, the invocation of Glanvill, much like the invocation of Bacon, Thoreau, and George Santayana, is in a context of steps "for mankind," a gesture of affectionate care.

He wrote for *presence*. By that word, or other words he was likelier to use for the idea, he does not mean visual or metaphysical presence of the god, nor does he presume absence of the god. He means, roughly, existential distance partially erased, but never nullified, by a sort of symbolic relevance, a consciousness of consciousness that has the quality of being consequential and reinforcing to his own. A childhood surprise, recalled in "Man Against the Universe," can illustrate this: "Perhaps my first awareness of the otherness of nature [was a] marvel enough—that a shell, a shell shaped in the seas' depths, should, without intent, so concentrate the essence of the world as to bring its absent images before me" (*Star*, 175). Absent it was, yet a new acquist of awareness, partly self-awareness but not exclusively: within his pages, man "stands and listens with a shell pressed to his ear." The essay in one sense embodies the surviving child, in another sense creates the brother. It reaches both inward and outward. It represents speech to self but not exclusively: it invites other talkers to self to partake and participate. In a metaphorically and psychologically more complex fashion, the essay may resemble archaeological fieldwork in being able "to read time from . . . the humanly touched thing [which is now gone yet which may yield] a kind of pity that comes with time" (*Night*, 84). For "nothing is more brutally savage than the man who is not aware he is a shadow" (*Night*, 85). The essay invites a kind of sisterhood and brotherhood in *pity* and affords a quasi-parental illumination silently evidencing one's own derivative status.

The essayist Eiseley came to this from having been a nonessayist who lacked exactly such symbolic consciousness, and knew it: "I was an outsider to whom the nail [in a Pawnee medicine bundle] could never denote more than a nail, or the flaked weapon [ice-age spearpoint] stand for more than a bygone historical moment. I was *afflicted* by causality, by technological time rather than the magic of *genuine* earth time"

(*Inv.*, 110; my emphasis). This, in its self-disclaiming, is an ironic undoing of an alternative mode of discourse, not the only or necessary mode
for science, but a mode that science at times employs abusively.

Eiseley is likewise ambivalent about presence, regularly flinching
from it after regularly proffering his consciousness to an audience imagined as so many individual fellow consciousnesses. It is as if he conceived the essay as Ortega conceived philosophy: as "not a science but,
if you like, an indecency, since it consists in baring things and oneself,
stripping them to stark nakedness—to what they are and I am—and
that is all." [19] Personal material may be interesting, Eiseley says in *All the
Strange Hours,* "*but* only if one is utterly, nakedly honest" (see above,
p. 249; my emphasis).

Earlier, he had acknowledged a colleague's note of praise for *The Unexpected Universe* and remarked, "I think it is sometimes painful to
write a book as personal as this one and expose it to critical attention." [20] Part of the exposure, again and again, is admission of habits of
disguise, metaphorized precisely as garments, whether the protective
coloration of sober academic or (doubly metaphoric) animal. His return
to poetry after thirty years apparently sharpened his ambivalence and
clarified it to him. He replied to one fan: "*The Innocent Assassins* was
largely composed during a long severe bout of viral pneumonia with
bad after effects which made it easier for me to prop myself up in bed
and write poems than to engage in the research necessary for a non-
fiction book of prose. They were a great emotional release to me and
now I find it is a little difficult to turn back to the world of prose." [21]
Some days later he complained to his medical friend Dale Coman of
feeling exposed and naked, with a desire to "disappear into the hills"
and to hear no more about it. [22] Earlier that year, soon after the publication of *Notes of an Alchemist,* he had initiated and apparently termi-
nated a correspondence with a Chattanooga reviewer: "As you will realize, there is something traumatic about publishing a book of verse at
sixty-five. In other words, in spite of vanishing with the snow leopard,

19. Ortega y Gasset, *Man and People,* 99.

20. Loren Eiseley to Philip Reiff (professor of sociology at the University of Pennsylvania), December 4, 1967, in University of Pennsylvania Archives.

21. Loren Eiseley to Kathleen H. Davison, September 11, 1973, in University of Pennsylvania Archives.

22. Loren Eiseley to Dr. Dale Coman, September 19, 1973, in University of Pennsylvania Archives.

the snow leopard could not resist peeking out from behind a stone. You can, however, consider me as having disappeared again, though this time not without a friendly purr directed toward you."[23] *In spite of, however, though not:* it scarcely requires the snow-leopard disguise in the passage to suggest a cat's-cradle of conflicting impulses. As a cluster of Derridian passages might gloss them: "Presence . . . at the same time desired and feared"; "One cannot help wishing to master absence and yet we must always let go"; "Without the possibility of *différance,* the desire of presence as such would not find its breathing space."[24] Eiseley aimed in the essays, as he did in much more limited fashion in the poems, for a meeting of minds and a brotherly accord of spirit, but not, though he had played football in high school, for anything like the round of hugs and hand slaps after a touchdown. And always not quite yet: there are a systole and diastole of approach and withdrawal, concurrence and demurrage. Time may recurrently seem a wound, or a deathly rite, the world a space of alien life or desert grave rather than a "green world . . . sacred center" (*Inv.,* 1). Brilliant scientific discovery and genius invention such as the zero (*Inv.,* 86; see *Ano.,* 23) may, although necessary, foster the abstractive death that is mere mechanistic causality, and give an entrance to "technological time" by suggesting a rationalistic extrapolation of a reductive future from an overmechanized present. But Eiseley, the self-defined fugitive, knows and abhors more the "sure intellectual impoverishment and opportunism which flight and anonymity so readily induce" (*Night,* 74).

So each essay, probably with some debt to Conrad's aspiration "above all to make you *see,*"[25] attempts to make present an emblematic moment of the "magic of earth time," what even the apposite Ortega put somewhat more spatially as the "lived-in forest. . . . What does not

23. Loren Eiseley to George Scarbrough, February 19, 1973, in University of Pennsylvania Archives.

24. Derrida, *Of Grammatology,* 155, 142, 143. Eiseley's notion of presence, as I am arguing, is less contingent on divinity than Derrida's. The latter's *différance,* fom *différer,* includes roughly *both* spatial difference and temporal deferral.

25. Joseph Conrad, "The Nigger of the *Narcissus,*" Preface. Leslie E. Gerber and Margaret McFadden rightly speak of several items in *The Night Country* and *The Star Thrower* as "owing much to the legacy of Twain, Bierce, and Poe," the "first-person tall tale" (*Loren Eiseley* [New York, 1983], 29), and more generally, of "Heine, Lamb, Hunt, Hazlitt, Emerson, and Thoreau" (p. 21). Eiseley also read August Derleth and H. P. Lovecraft. But I think the example of Conrad is more germane. He owned Edward Said's *Joseph Conrad and the Fictions of Autobiography* (Cambridge, Mass., 1966).

exist nor have sense is the forest 'in itself,' neither 'in itself'—realism—
nor 'in me'—idealism."[26] The forest in any case is limited, or the maker
or making is limited. No one so death-haunted as Eiseley would claim
boundless potentiality for the reflective, imaginative mind in any individual, whatever might be a reasonable hope for the species or life-
forms beyond present temporary configurations of the species, and
whatever might be Eiseley's own care for his crafted words: "One has
just so many pictures in one's head which, after one has stared at them
long enough, make a story or an essay" (*All,* 156). His essays rest on a
meditative composition of place—but not quite on Ignatian meditation.

"Anecdote," he wrote, "allowed gently to bring under *observation*
thoughts of a more purely scientific nature" (see above, p. 249; my emphasis). Given that "anecdote" is as likely to close as to open or center
the Eiseleyan essay, those words sound like an ironically if not disin-
genuously modest redaction of his earlier praise for the man who was
almost his secret sharer, Francis Bacon: "It can of course, be argued that
without some hint or idea of what we seek, fact-gathering will take us
nowhere. Bacon . . . was aware of the way the mind runs in ascending
and descending order from fact to generalization and back again, 'from
experiments to the invention of causes, and descending from causes to
the invention of new experiments. . . . Knowledge . . . hath much
greater life for practise when the discourse attendeth upon the example,
than when the example attendeth upon the discourse'" (*Man,* 54). *Example,* obviously enough, better names the typical Eiseleyan personal
"picture" than does *anecdote* or *experiment.* But the "ascending and
descending order" again and again characterizes the progression of his
essays.

The "more purely scientific" thoughts or ideas vary somewhat at the
level of intermediate generality. They may be biological, of evolving
forms or persisting kinships across form or about partially explicable
but quite unpredictable trains of cause and effect in nature. Such biological thoughts frequently parallel archaeological and anthropological
ideas that are analogously attentive to cultural evolution and devolution, or to persistence. The feature of discovery or interpretation that

26. José Ortega y Gasset, *Meditations on Quixote* (New York, 1961), 175. *Not* to the
contrary: The tree at the climax of "The Brown Wasps"—which turns out to have been an
imaginary growth from a lost seedling—because it "somehow stood for my father and the
love I bore him," has "bloomed on in my individual mind, unblemished as my father's
words" (*Night,* 235–36). It is a lived-in miniforest.

links the Eiseleyan focuses of biological or cultural persistence, on the one hand, and the "break" in biological or cultural or individual being, on the other hand, is unexpectedness. The words *the unexpected universe* might have entitled any of his other collections as well, including his most overtly autobiographical book; its title, *All the Strange Hours,* suggests that he might have agreed.

The fact of unexpectedness—whether in genetic adaptation or improbability, cultural or individual oddity or shift, or surprising insight—is a more general kind of fact, as Eiseley emphasizes assertively in virtually all the essays. Insofar as the reader trusts him, that might be enough. But he seems to have been diffident about the level of trust he would be accorded; he tells enough stories of academic and other rebuffs to indicate tenderness on the point. In any case, he quasi-dramatizes the unexpectedness of the universe with the unexpectedness of the essay.

Within a subsection, one quickly comes to expect a *scene* and rather more framing of scientific or otherwise general interpretation and reflection. Occasionally a subsection will have two scenes. All tend to be starkly visual—although one reads occasionally of sound effects (a bird cry, a wisteria pod's exploding)—and a few are doubly visual in reconstructing a representation Eiseley had read (Homer's, say, or Thoreau's). But the content and sequence among subsections in essays with more than one, like the content and sequence of essays in any of the collections after *Immense Journey* (1957) and *Darwin's Century* (1958), is increasingly unpredictable, except, to a degree, in *All the Strange Hours: The Excavation of a Life* (1975), where chronology tends to govern, as it did in *Immense Journey* and *Darwin's Century.*

Of course, total unpredictability would mean total unintelligibility, and Eiseley's unpredictability quickly becomes one of several shaping constants. The reader may ask "Where will he take us now?" But the question will not envision possibilities ranging from heaven to hell. Eiseley gives grounds for expecting a writerly presence often more *explicitly* alone than that of many an earlier essayist, even when he presents to us through the one-way filter of his text some transaction with other texts or persons or beings. The relation he reports may vary from antipathy and rejection through ambivalence to affection and even introjective intimacy (as with blood cells or an old gull or a derelict passerby).[27]

27. See "The Inner Galaxy," in *Unex.,* and "One Night's Dying," in *Night.*

Eiseley's reader has come to expect that the general comment and interpretation will evince a relationship to its audience ranging over a markedly lesser range: deliberation shading into reflection, sobriety sometimes relishing happy occasions but much more often somber. Typically,
and certainly at the essays' best, they keep reportorial hold on the occasion, do not relinquish the urgently communicated sense of, I was there,
altogether there.

Second, it can be expected that the ranges of paleonaturalistic and
cultural matter and reflection will be punctuated with metascientific
precept and example on two heads. How shall science function? Not
callously with regard to persons or other creatures or the web of nature.
For what shall science function? Not for knowledge, as some would say,
even less for any narrow technocratic vision, but, in the broadest sense,
"for the uses of life." Encounter with that phrase of Bacon's from *The
Advancement of Learning* seems to have been one of the crucial moments in Eiseley's intellectual formation (*Man*, 63).

Concomitant with exhibiting the unexpectedness of the universe and
situating science in a biophilic context, Eiseley regularizes his essays in a
third way, though the reader may become aware of this less quickly because it is less insistently explicit. It is the grounding of the properly human life world in responsibility. Eiseley can profess the linkage between
unexpectability and responsibility: "Modern man, who has not contemplated his otherness, the multiplicity of other possible men who
dwell or might have dwelt in him, has not realized the full terror and
responsibility of existence" ("Strangeness in the Proportion," in *Man*,
115–16, and *Night*, 148).

Similarly, his twist of one of the most pervasive and fruitful metaphors in the Western tradition, the journey, occludes the linearity and
highlights the contingency: "Upon journeys . . . man in the role of the
stranger must constantly confront reality and decide his pathway"
("The Chresmologue" [a piece he had variously called "Man, Time, and
Prophecy," "Man, Time, and Contemplation," and "The Scientist as
Prophet"], in *Night*, 62). The point is important: when it is clear that
even something suggestive of linearity—as a journey is—is for him emphatically not deterministic, it becomes easier to credit the pervasiveness
of his doubt that anything is deterministic. Book dedications to his forebears and oblique references scattered in the autobiographical writings
reveal the family responsibilities he accepted. The unexpectedness of the
structuring and sequencing in his essays needs also to be understood as

choice he self-consciously exhibited. At some point he jotted down but evidently discarded a tentative subtitle for *The Unexpected Universe:* "The Odyssey of a Scientist."[28] Odysseus endured the caprices of the gods, but Eiseley would not put it so.

His titles are significant. *The Immense Journey* could be the title of a linear account of evolution, but the chapters could stand alone as essays and most of them had done so during the previous decade. Most of them are like most of his essays, titled with a noun phrase suggestive of a label on one of the pictures he meditated. The time line wavers; "digressions" appear; contingency is emphasized. A few of his titles suggest a complex tableau, fewer still an argument more or less historical: X is (or is not) or did or became Y. But most of his titles suggest an act of textual definition, with some existential complexity of argument; an appropriate subtitle of length appropriate to his Renaissance interests and faithful to his position and procedure might take the form, Wherein, being accorded trustworthy presence in the reader's consciousness, the author demonstrates how to see X as Y, in reproach to false gods of totalizing objectivity.

In the early and none too fortunately entitled *The Firmament of Time,* the component essays start unrolling with the mildly cryptic but commonplace linearity of chronology and logic suggested by their titles: "How the World Became Natural," "How Death Became Natural," "How Life Became Natural," and "How Man Became Natural." But then "How Human is Man?" and "How Natural is 'Natural'?" All, all is problematized!

His titles acknowledge his growing preoccupation with intellectual apprehensions of a kind infringing positivism's canons, and with moral responsibility, in a universe in which epistemological and other boundaries can be as unexpected and problematical as anything else. "The Invisible Island" (in *Unex.*), whose title is redolent of oxymoron, turns out to concern the indeed paradoxical island, boundless yet constraining, of language and culture. The 1970 essay collection *The Invisible Pyramid* refers to what is more obviously man-made—and *Another Kind of Autumn,* his late collection of poems, similarly canvasses linguistic and cultural boundaries.

He is explicit about indeterminacy, the fourth constant feature of the unexpected essay, in "The Chresmologue" (in *Night*), a fair account,

28. University of Pennsylvania Archives.

over and above a forgivable amount of special pleading, of the poetics of his kind of essay. There he refers approvingly to Henri Bergson's exposition of life's indeterminacy and to Karl Heim's notion of the present as being "still in the molten phase of becoming" (*Night*, 63, 71). The chresmologue he explicitly takes to be the prophet of saving reality to a culture that in ignorance or through the worse agency of bad mythology plunges like a runaway coach toward the abyss. He endorses Ralph Waldo Emerson's contention that nature "is a mutable cloud." A man of hidings, he pronounces the "human future" that will be made of ourselves to be the "sure intellectual impoverishment and opportunism which flight and anonymity so readily induce" if "our thought runs solely . . . upon the clever vehicles of science" (*Night*, 74).

He writes and reiterates that "in the world, there is nothing to explain the world" (*All*, 238, 242; as quasi-disguised quotation in "Science and the Sense of the Holy," in *Star*, 158). But, he demonstrates, one can talk about a mutable cloud and avoid sounding like Hamlet and Polonius' dialogue on clouds—can do so by a recourse in good faith to shared or genuinely shareable systematic knowledge.

And Eiseley describes man in the world the chresmologue faces as "an indecipherable palimpsest, a walking document initialed and obscured by the scrawled testimony of a hundred ages. Across his features and written into the very texture of his bones are the half-effaced signatures of what he has been, of what he is, or of what he may become" (*Night*, 59–60). Accordingly, the chresmologic essay is a writing of reading, of deciphering, of teasing out subtexts in a palimpsest. It is a writing *of* reading in both senses familiar to every writer contemplating a complex text: the writing *represents* thoughts attendant on reading; and the writing furthers—even creates—and *presents* thoughts occasioned by reading, including thoughts of puzzlement. Reading and writing are step-by-step linear, like a journey afoot. Literizing and journeying are alike, in Eiseley's usage, not least in being sometimes impelled by needs or apprehensions of which one is not conscious. Writing, reading, journeying are all three subject (how well we know) to involuntary pauses, impulsive flights, deliberate side trips—which thereupon for better or worse become part of the reading, writing, or journey.

The chresmologue or, as he can call himself, the shaman mediates across, and thereby dislodges or redraws, seemingly firm boundaries, as between the animal and human, the evident and the hidden, the fixed and the unfixed. Such mediation is a fact, a quasi fact, and a metaphor.

It is the presentation of a way of being in time that is radically alter-native to the awful clatter of tychastic time. In "The Inner Galaxy" (*Unex.*, Chapter 8), Eiseley conflates "personal anecdotes"—about his care-filled approach to an observatory and the cold rebuff he received on behalf of cash register and casual viewers—with subsequent address from his "inner sky" (which can be an "interior void"), with another about a fall and laceration's spilling creaturely corpuscles from his inner physiological galaxy, with still another about fellow-creaturely con-templation of an old gull. The conflation warrants defining evolution as love for life within life. He had considered using an Emersonian phrase to entitle the piece "The Bright Stranger: A History of Love."

Thus, Eiseley concurrently emphasizes responsible choice in the read-ings, writings, and journeys undertaken in an open-ended natural and human universe. Sometimes responsible choice in indeterminate or mys-terious worlds entails suspending judgment and continuing the search for elusive truths. The humanoid way of technocratic society was em-blematized for him in a "terrible" way by a "dead-faced derelict" on the train out of New York who roused himself just enough to proffer money and ask the conductor for "a ticket to wherever it is" (*Night*, 63). The terribleness Eiseley felt and his sense that the man "had personalized the terror of an open-ended universe" were of course in some measure pro-jections of his own anxiety about where he was going in life and in death. But he wrote for readers similarly incomplete and hence anxious in their assurance. His own uncertainties shed light, too, on the frequent ending of the chresmologic essay in a somewhat inconclusive fadeaway—as if to say, Here is my sibylline reading, my disentangled bit of pa-limpsest, my prophetic remap of the often-misconstrued itinerary; take it or leave it, but I must go in search of more. Contemplating "The Ex-cavation of a Life" in galley proof[29] (the phrase is subtitle to *All the Strange Hours*), he added to his own palimpsest, "Ironically, I who pro-fess no religion find the whole of my life a religious pilgrimage" (*All*, 141). The essayist so unfixed, unsituated, appropriately presents a kind of dialectic of—in the Derridian pair—protention and retention, doing and undoing. "One is conscious of an always shifting point of view which is a true part of the fragmented world we occupy. The message derived from the fragments cannot be the message from the whole":

---

29. So called in the corrected copy in the Heritage Room Collection, Bennett Martin Public Library, Lincoln, Nebr.

these words of Eiseley, ostensibly about David Lindsay's *The Voyage to Arcturus*,[30] apply to the ambivalent views of world and humankind, and to the uneasily companionable views of readership, that Eiseley took in his own essays, which can confess to occasional disguise.

His views of the world—variously as a mysterious web of being and an antagonistically alien "outside" (and indeed partly an antagonistic inside)—relate crucially to his views of and postures with respect to readers. One may well conclude that he agreed with the Ortegan point whether or not he read the page that laid out a "great paradox: it is not *the* unique and objective world that makes it possible for me to co-exist with other men, but, on the contrary, it is my sociality or social relation with other men that makes possible the appearance, between them and me, of *something like* a common and objective world."[31] It seems central, upon reflective rereading, that "personal anecdote" and "personal material" do a good deal more, and with more complexity, than "bring under observation thoughts of a more purely scientific nature." The anecdotes, variously narrative and emblematic, and the rhetorical resituatings of self and readership relate somewhat as medium and message. Part of the fascination of Eiseley's essays is the frequent ambiguity whether the scientific or the personal is the medium, the other being the message; it is the ambiguity as to which—in alternative terminology—is the figure and which is the ground. A "history of love" it is indeed, even if Eiseley, not uncharacteristically, suppressed that phrase as a title.

A catalog of the modes of Eiseley's writerly sociality would include all those voices which in his essays make possible between himself and his readers *something like* Ortega's common world. There is, time and

30. Loren Eiseley, Introduction to *The Voyage to Arcturus*, by David Lindsay (New York, 1963), vii. The book is a republication of the first edition, of 1920. Eiseley calls Lindsay's tale a "story of the most dangerous journey in the world," the journey "into the self and beyond the self" (p. viii). He verges on the defensive in saying that Lindsay "was perhaps too honest to record one voice alone among the many conflicting voices that represent the living world" (p. x). The "perhaps" is Eiseley plain. Concerning Eiseley's "fragmented": E. Fred Carlisle, writing about the bleakness in early poems on ruins (see *Loren Eiseley: The Development of a Writer* [Urbana, Ill., 1983], esp. 57–62), has prompted me to think about the doubleness of fragments and ruins in Eiseley's mature prose. They are bleak but are also a medium for new human life, in that human life is a matter of cultural, as well as biological, evolution. For another version of such doubleness, see Jean Starobinski, *The Invention of Liberty* (New York, 1979), 115.

31. Ortega y Gasset, *Man and People*, 108.

again, the prophet-nostalgist of the green world and of childlike innocence, eager to reinforce the nostalgist in any of us. At the end of *The
Invisible Pyramid,* in "The Last Magician," that poignant stance can resemble Nick Carraway's at the end of *The Great Gatsby:* nostalgia can
shade into regression. But when it does, the reader is situated not as a
fellow-in-feeling but rather as a witness to testimony, perhaps fellow to
a backgrounded element of the writerly self. For example, Eiseley
writes, "*I* loved the darkness. *I* feared it yet returned to it. It was the
mother out of which I came" ("The Places Below," in *Night,* 27; my
emphasis), as if to say, You may be otherwise, but note this human possibility, as I note it.

Obviously and frequently, he is the prophet-sentry, warning that our
situation is not as it has been or as we might have been lulled into thinking it is. "The Hidden Teacher," in *The Unexpected Universe,* is a good
example. He would teach us, we presumably are learners, but the sentry
is one of us, even if situated apart. In an occasional meliorative variation, the sentry becomes, prophetlike, incipiently programmatic. But
instead of crying, "Repent!" to "anticipatory man" (see "The World
Eaters," in *Inv.,* 68–69), he characteristically writes, "*Men* should discover their past. . . . Only so can we learn *our* limitations and come in
time *to suffer* life with compassion" ("The Crevice and the Eye," in *All,*
85; my emphasis). This prophet is apart, but one of us, among fellow
subjects. His is the position and role of plainsman as good neighbor.

The sentry can occasionally view his subject with the closeup eye and
existential middle distance of amusement, more or less uneasy, more or
less self-directed, as in "The Creature from the Marsh" (referring to
himself, and "all of us"; in *Night*) or "The Obituary of a Bone Hunter"
(referring to himself ironically dead, because implicitly reincarnated as
essayist prophet and presence; in *Night*). A separate essay could be written on Eiseley as an American humorist; it would be a short essay.

With a more startling degree of distance, sometimes involving a met-
amorphosis so abrupt that it may have been suggested to him partly by
film technique, the sentry may become an alien observer. The distance is
the idiosyncratic thing; after all, Nick Carraway described himself as
both inside and outside the life around him, and many another literary
narrator or expositor might have done so. But few or none have been so
far outside, simultaneously, as Eiseley. He glosses the matter in "Days of
a Thinker" (*All,* Chapter Eleven). A cave experience early, it seems, in
his professional life had left him feeling a temporary parolee "from

some kind of indefinable death" and seeing his own hand as "ten thousand years away. So were my eyes, so *would they always be.* . . . 'Dwarfed,' I said . . . seeing *ourselves* moving with tiny gesticulations across an infinite ice field . . . like a glimpse through the slitted bone with which Eskimos protect their eyes from snow blindness. I have never had occasion in the years since to think upon *us* differently. Not once" (*All,* 104–105; my emphasis).

It should become increasingly apparent that the self-containment of that autobiographical chapter is thematically as well as temporally compromised. After the cave experience, he went on living and writing (although scarcely so after *writing* that experience), and not always as if a dwarf among dwarfs. Does one bother to write "of the last things" ("The Inner Galaxy," in *Unex.,* 176) for "a little lost century, a toy," and to address the like to the "modern world . . . constricted beyond belief" unless a possibility of transcendence is apprehensible? The end can be apocalyptic, the eschatological Eiseley may note (for example, in "Man in the Autumn Light," in *Inv.*), yet writing of the last things may be not so much apocalyptic as faithful to a hope of transcendence. One of his most anthropologically and autobiographically provocative interpretations proposes that "perhaps cultures, like individual men, are destined always to betray themselves and then endlessly to relive their bereavement" (*The Shape of Likelihood,* Introduction, 17). The *perhaps* might make us fellow interpreters; as such, we might be interested in his more pointed adjuration that "one eternally keeps an appointment with one's self" (*All,* 169).

Despite scattered instants of self-overdramatization and even unfocused pain, despite the ready acknowledgment of vast lifelessness and hostility to life in the universe at large and the human world within, despite the constant stand in opposition to our culture of taxonomized spatiality and mechanized temporality, Eiseley's dominant emphasis is upon a kind of argument he makes mostly by sympathetic report and witness for transcendence of "the prison" man is—for transcendence through love. "The Innocent Fox" and "Inner Galaxy" (in *Unex.;* reprinted in *The Star Thrower*) and "One Night's Dying" (in *Night*) exhibit this. A late poem speaks of being momentarily "outside this inside of nature" ("Five Men from the Great Sciences," in *Inno.,* 72–73), but the more substantial, structured, and indeed pervasive essence of the Eiseleyan essay is the appeal to love, to whatever capability the reader has—given a bit of briefing and encouragement—for loving the greater and lesser beings of the world and the "lost ones," surely including Loren Eiseley.

The evidence reposes in the University of Pennsylvania Archives that many, many read his essays as, in some sort, calls to comradeship-in-arms or to gather around the campfire or in study.[32] Scores or hundreds of readers wrote to him in that vein. Two remarkable features of the correspondence are germane here: He seems to have answered all those letters. And the answers nearly all exhibit a kind of stilted courtliness contrasting with the familiarity and even intimacy he evidently felt he could venture with the quasi-sibling Other he addressed in his essays.[33] That somewhat idealized Other is of course ourselves, and perhaps we are flattered. He acknowledges that "man is not one public, he is many" ("Man Against the Universe," in *Star*, 167), but he writes primarily to those already sufficiently ill at ease in Zion to be likely recruits to his cohort, and comrades-in-arms.

A further definition of the Eiseleyan essay, as formally an *engagement*, appears in a few words he selects from Francis Bacon's *pluribus*, in "The Hidden World": "The unlearned man knows not what it is to descend into himself or call himself to account . . . whereas with the learned man it fares otherwise, that he doth ever intermix the correction and amendment of his mind with the use and employment thereof" (*Man*, 65; from Bacon's *The Advancement of Learning*). That the Eiseleyan mind has been stocked and trained and used is obvious enough to need no special announcing.[34] But without Eiseley's characteristically veiled hint, it might not be quickly noticed how frequently he descends into himself, sometimes to report, sometimes to exhibit, a correction or amendment of mind.[35] Christianity is sometimes character-

32. Carlisle has implied that the *poems* are loving acts (*Loren Eiseley*, Chapter 5).

33. A *very* few letters to old friends are unguarded. There is graceful correspondence with an elderly but lively-minded lady in a nursing home. To instruct about reviewing and the publishing game, Eiseley plays a gruffly affectionate Frank Speck to Annie Dillard's (so to speak) Lorena Eiseley. There is little else.

34. With a Ph.D., Eiseley was in 1937 "already an informed and mature anthropologist with a systematic and comprehensive grasp" (Carlisle, *Loren Eiseley*, 108).

35. Eiseley had "long shared Van Wyck Brooks's view that American critics have not paid proper attention to the nature essay. My own feeling is that [its] origin extends further back than McGehee seems to think" (Loren Eiseley to Don Federman, May 20, 1974, in University of Pennsylvania Archives). See also Loren Eiseley to Dr. Warren Weaver, September 23, 1974, in University of Pennsylvania Archives. For Brooks, see his *From the Shadow of the Mountain* (New York, 1961).

ized as "good news, not good advice." This self-professed lifelong pilgrim who could "only seek"[36] tends to write guardedly encouraging news less translatable as advice than as model or witness.

But someone looking for clear news or direct advice will begin to suspect that the scene in Eiseley's essays, the forces in their discursive world, would disorient almost any compass. Eiseley's metaphors may enter a second order, as when the familiar one in which the terrestrial journey stands for the individual life and for individual or corporate pilgrimage toward grace becomes a complex signifier of biological evolution or Eiseleyan journey toward . . . what? That does not show, in Wallace Stevens' phrase, glossed by Louis Martz as a definition of the meditative poem, the "mind in the act of finding what will suffice."[37] The Eiseleyan essay offers a privileged moment, and sometimes a celebrative instant, but rarely the finding of what will suffice. Its findings, insofar as they are of nature, are vulnerable and fated, and insofar as they are of the mind, they are for him almost invariably ambiguous. So it is with science and, I think, with the metonymically implicated and metaphorically implied scholarship of the humanities. So it is with the mind itself, "sure to question . . . and to be at me day by day with its heresies until I grew to doubt the meaning of what I had seen. Eventually darkness and subtleties would ring me round once more" ("The Judgment of the Birds," in *Imm.*, 176; also in *Star*). It would be equally fair to see these essays as prose poems of the mind in the act of divesting what will not suffice, including not only the false but also unwarranted degrees of certitude. Both knowledge and certainty imply considerable order. But order "is at least partially an illusion . . . of a time, and . . . fraction of the cosmos, seen by the limited sense of a finite creature" ("The Unexpected Universe," in *Unex.*, 46). In a general exposition of postmodernist and poststructuralist theory, Edward Said has elegantly

36. Loren Eiseley to Dr. Bolton Davidheiser, January 16, 1967, in University of Pennsylvania Archives. See also Loren Eiseley to Randolph G. Adams, November 14, 1949, in University of Pennsylvania Archives.

37. See Louis Martz, "The World as Meditation," reprinted in his *The Poem of the Mind: Essays on Poetry, English and American* (New York, 1966).

summarized the situation: "And since there is no foreordained, a priori route for the discourse to travel, a route legislated from outside itself, the order of statement, discourse, and *épistémè* bears no resemblance to the traditionally concentric circles of self, world, and God, which are held together dynastically . . . by three types of continuity, all . . . guaranteed by God the Father. The order of discourse is maintained by *legislated accident,* by chance: 'The present is a throw of the dice.'"[38] He could almost be speaking as well of Eiseley as of Foucault. Eiseley, the acknowledger of chance and accident so grand as to seem fateful, wrote very late to a friend, "I have come to believe with Spengler, that one can only try to endure with a certain stoicism what one sees as fated, and give all oneself and pass on what little of love and sense of duty one small ephemeral body may possess."[39] But what Eiseley passes on, the reflected-upon scenes from the palimpsest of his own nature and from other men's texts and from the book of creature, he will "strew . . . like blue plums" (*Unex.,* 232). As significant events in his life, they were more or less accidental—certainly unpredictable—in a pattern similar to the pattern of their appearance in the essays. Which is signifier, which signified? Each is a metaphor of the other. Moreover, the plums, falling in so unpredictable a pattern, even when partially organized by autobiographical chronology, are at a distance from the tree.

Eiseley does not explicitly sanction that last extension of his figure. But he appreciated the fact too well not to know that, as Louis Marin has put it, "'Story-telling' writing creates the simulation of an absence; it creates in saying what is said a semblance of silence."[40] "Infinite love" may be the "solitary key to the prison that is man" (*Inv.,* 41), but the distance entailed by passing the love on through print is partly something Eiseley desired. No essay would be his without its consortium with his sense of dwarfed race and fleeting slime-mold cities and the felt abrasiveness coming from both outside and inside his own consciousness.

38. Edward Said, *Beginnings: Intention and Method* (Baltimore, 1975), 310–11.

39. Loren Eiseley to Frederick R. Abrams, January 3, 1974, in University of Pennsylvania Archives.

40. Louis Marin, "Writing History with the Sun King: The Traps of Narrative," in *On Signs,* ed. Marshall Blonsky (Baltimore, 1985), 267. Also germane are Walter Benjamin's reflections on the storyteller's abbreviations and shifts since the idea of eternity and the presence of death have faded. See "The Storyteller," in Benjamin's *Illuminations,* trans. Harry Zohn (New York, 1969). But death was no fading presence for Eiseley.

Nevertheless, for all his conflicts, which give substance and conflictual edge to his essays, there is an abiding trust in the memory that enables and structures them, precisely because it is not mechanical. (Admittedly, some readers will wish that it structured them more.) Memory loses much but, Eiseley believed, saves what is important; or if it saves unpredictably, one can reflectively find importance in that, a junk cart at Fourteenth and R streets, or whatever—it comes to the same thing essayistically even if not logocentrically. And print, not so much a sinister mechanical thing for Eiseley the bibliophile as it is admired human craftsmanship—print will preserve the fruits of memory a bit longer on "heartbreak earth, so loved, so multitudinously/extinct" (*Ano.*, 58).

The other conception animating the interplay of remembrance, approbatory exposition, adversarial exposition, and reflection in these essays remarkably corresponds to Ortega's notion of *compresence*. Speaking of the being of the other person, Ortega says: "This inwardness or intimacy . . . is never present, but is compresent, like the side of the apple that we do not see. And [it] is more amazing. For in the case of the apple the part of it that is hidden at any moment has been present to me at other times; but the inwardness that the other man *is* has never made itself, nor can ever make itself, present to me. And yet, I find it there."[41]

"And yet, I find it there." No remark better accords with Eiseley's practice, which could as appropriately claim the title *Findings and Keepings* as his colleague Mumford's autobiographical volume bearing that title. But there are two signatural variations. Compresent to Eiseley are not only various other men (and very recessively his wife, and nameless girls in his early poems and later essays) but birds, a dog, a fox cub, a cat, and orb spiders. Second, he, more anxiously than most, would be compresent to us. He puts explicitly in a poem what is implicit throughout the essays, adjuring an unnamed "beloved" to "seek for lost things and strive to know" what is "behind the mask you scan" ("Deep in the Red Buttes," in *Wings*, 33). She scans, the formal supposition is, both his face and the words of his poem, for an inner him.

The reader scans his essays and finds in this unorthodox pilgrim something like the "theology of juxtaposition" his sometime fellow Philadelphian Harvey Cox advocates—a "method for theological jesters. It questions not only the self-evidence of the tradition but also

41. José Ortega y Gasset, *Man and People*, trans. Willard Trask (New York, 1957), 92.

the self-evidence of experience. It challenges the past from the perspective of present experience, and it challenges the present from the perspective of our memory of the past. It limits the claims of both past and present by thinking in light of hope [and] it focuses precisely on those discomfiting points where memory, hope, and experience contradict and challenge each other."[42] The Eiseley wryly conscious of the folklore figure of the mocking trickster was himself a jester of the dry mock, an ironist at pretensions to totalizing adequacy promulgated by any temple, whether scientific, academic, economic, professional, or ecclesiastical. Nor is Cox's term *theology* any more out of place here, given Eiseley's sustained questioning of that in which he could, and we might, place our ultimate trust. Given his more than curiosity about whatever might by definition exceed definition, "discomfiting points where memory, hope, and experience contradict and challenge each other" suggests as a descriptive label the facts of adjacency and parts in these essays, their metonymic element, which in turn is a metaphor of contemporary conflicted disparateness. The personal anecdotes are clearly offered as emblems of compresence, as synecdochic. And one must acknowledge the relevance for Eiseley of Walter Benjamin's point that "death is the sanction of all that the storyteller can tell." But is there a master metaphor, a second-order sign to which all metaphoric and metonymic and synecdochic signifiers refer? It is possible to take the Eiseleyan essay as a great as-if, a metaphor of *is* with ironically indirect or sparing hints of what may be, namely, the moment of companionship in the interrelatedness of life, life understood to be mysterious, not merely puzzling. Eiseley comments on seeing a row of telephone booths next to a cemetery in a "bleak countryside" (*Unex.,* 45). Like Mumford and Tate, he has watched and meditated in the darkness. Like them, he has called from the darkness, experienced bad connections, yet remained hopeful that some of us passing the cemetery will answer the phone.

42. Harvey Cox, *The Feast of Fools: A Theological Essay on Festivity and Fantasy* (Cambridge, Mass., 1969), 133. Though not the *succès fou* of his earlier *The Secular City,* this was widely reviewed, the more so in the Philadelphia area. Eiseley might well have seen the book in shops or in the Van Pelt Library at the University of Pennsylvania.

9

PROPHETS AT THE TURBULENT BOUNDARY 🌿

Come to me, stand with me! From far horizons
Let the light be brought to pierce the darkness.
My brother, O my brother of the distant generations,
To you I cry out from the bosom of the dark.

—Shin Shalom, "Strength, Son of Wonder"

I n the sense in which Mumford, Tate, and Eiseley are prophets, it can be said that the direst jeremiad of the prophet is, in its assumptions, a cry of hope. Not in any proverbial sense that the night is darkest just before the dawn, but in an anti-Manichaean sense that only because light exists is darkness visible; light is from all time privileged. Less metaphorically: life has occurred and will in the large seek life, not death.

A splendid essay by Jean Starobinski, recently reprinted, makes of Jean Jacques Rousseau a conveniently clear and relevant figure for definition by partial contrast.[1] (In what follows, I will use *Rousseau* as shorthand for *Starobinski's Rousseau,* sidestepping questions of the adequacy of Starobinski's representation, questions I have neither need nor competence to address.) Rousseau, for Diderot the "Cato and Brutus of our age," was for Starobinski "perhaps the prophet of transition" from the sphere of religion to that of politics—either mode involving a sinner repenting and turning from his wickedness to live, atone, follow a different path.[2]

Mumford knew and in part admired Rousseau's work, even wishing, with reservations, to emulate his prose style. And Mumford, more even than Tate and far more than Eiseley, had in mind the model of the Ro-

---

1. Jean Starobinski, "The Accuser and the Accused," *Daedalus,* CXVII (1988), 345–70; originally in *Daedalus,* CVII (1978).
2. *Ibid.,* 345, 357.

man civic moralist. But that mode is most marked in a number of what may be called Mumford's tracts for the times,[3] not dealt with here, and in *The Condition of Man* and *The Conduct of Life*. *Emile* is more a how-to-do-it book for the betterment of the *polis* than anything I have considered by the three American essayists of the present volume.

Put positively, the three are less the political projectors and more, as well as more ambitiously, the representatives of a radically different world, different because of a somewhat demythicized ontology. Eiseley, it is to be recalled, added, "My life has been a religious pilgrimage" to galley proofs of *All the Strange Hours;* and his first book was *The Immense Journey*, about evolution. But even in the chapters of *The Immense Journey*, as in *The Unexpected Universe* and his other collections of essays both more and less sequenced, his *predominant* trope is not the pervasive Western metaphor of journey but a kind of synecdoche of new ontology and new community, interfolded with the old, new playscripts in the mysteriously given *theatrum mundi*. He never used "The Odyssey of a Scientist" as a subtitle, and he canceled "The Odyssean Voyage in Science and Literature" as the title to what became "The Ghost Continent," the first chapter of *The Unexpected Universe*. Tate, even before his reception as a Roman Catholic convert, seems like Mumford to have envisioned not just a linear turning from a lesser toward a profounder reality—although that would not be a grossly reductive description of their hopes. But for them it was less a linear turning than a turbulent *confluence*, that is, critical jostlings on the journey. For Tate, the turbulence is primarily of enlivening language fostered by the communal republic of letters; for Mumford, it is primarily the variously connected freemasonry of "organic" being and doing. From and within the turbulence, small changes may be expected to produce, in unpredictable ways, great changes. The three essayists may be seen to assimilate the biblical, Augustinian quest or the Vergilian path of duty— Mumford trolling through history for *explorations, progress, advancement;* Tate being still in order to know, as if on the "keel of a capsized boat" (*The Forlorn Demon*, Preface), mayhap to hear a symbolic buzz; Eiseley fleeing and questing, questing in part to assimilate flight to a renovated tradition of generative turbulence. In that line their forebears

3. See, for example, Lewis Mumford, *Men Must Act* (New York, 1939); Mumford, *Faith for Living* (New York, 1940); Mumford, *Values for Survival* (New York, 1946); and Mumford, *In the Name of Sanity* (New York, 1954).

include a voice from the whirlwind, Heraclitean fire and flow, and Lucretian metaphoric hydraulics.[4]

The prophetic vocation of the three essayists can also be approached through a revealing difference from Rousseau in their stance toward their imagined readership. Rousseau "finds it urgent to warn the civilized world against lethal dangers. . . . The initial gesture of accusation established a dramatic relationship between the writer and his contemporaries. . . . Accusatorial thought fosters antinomies, . . . is spontaneously 'Manichaean,' . . . [entails] a *living* antagonism."[5] This is recognizably germane to the three Americans, who deal not in abstract hypotheses or options but in concrete and often populous situations in which opposing persons may be stigmatized. But for them the "insensate" (to use what is almost a signature word with Mumford) are rarely a category or even a group; they are typically individuals completed by history or literature and redeemed as exempla, like Descartes. The living and the yet to live, however benighted, may abruptly or gradually awaken to a larger and better-illuminated world, the prophets' readers are typically given to understand. The universe is not, for the American three, Manichaean, although it may be a playhouse in which dark scenes can be enacted in a larger drama. And more generally and profoundly, these prophets are clearly imagining with some tincture of affection a potential readership of the kindred in spirit, persuaded or persuasible as to the argument, with regard to whom the immovable are an ontologically failed third party.

The difference continues consistently with regard to the register and orientation of reproval of ill. "Rousseau's disturbing innovation . . . consists in not making of this reproval the premise of an act of contrition. . . . A social ideal is openly preferred to a religious ideal. . . . [He] denounces like a theologian, but he denounces the *political* nature of evil . . . [on behalf of] *civisme* . . . a *substitutive sacred value*."[6] Of the very essence for the twentieth-century American prophet as represented by Mumford, Tate, and Eiseley is the dismantling of a *substitutive* sacred value. Its dismantling entails for them denunciations not of the political nature of evil but rather, and more profoundly, of the *phan-*

4. I urge this the more confidently thanks to Michel Serres' brilliant *Hermes: Literature, Science, Philosophy*, ed. Josué Harari and David F. Bell (Baltimore, 1982), esp. Preface, Chapter 9 (on Lucretius), Postface by Ilya Prigogine and Isabelle Stengers.

5. Starobinski, "The Accuser and the Accused," 346–47, 354.

6. *Ibid.*, 350–52.

tasmally and *globally* intellectual nature of a reigning *error* and, secondarily, of its evil and political consequences. They denounce or, better, decry in behalf of a consciously indeterminate and developmental relationship with others and a transcendent Other, *not* in confident relationship with a determinate and essentially immanent value.

What is the nearest thing to an American *civisme* in this corpus? Scattered passages in *The Condition of Man, The Conduct of Life,* and *I'll Take My Stand,* but *not* passages by Tate in the last of these. Moreover, the essayistic denunciations tend to enfold mea culpas or to be framed by mea culpas in letters published subsequently—most ambivalently and distantly in Mumford's case. No one of the three writers is altogether an outsider, as Starobinski's Rousseau repeatedly styles himself; rather, their characteristic stance was foreshadowed, as was noticed of Eiseley (see above, p. 265), by the self-description F. Scott Fitzgerald put in the mouth of *The Great Gatsby*'s Nick Carraway: "both inside and outside."

Moreover, Rousseau "has prefigured the radical proposition, which, by sacralizing politics, mixes the metaphysical expectation of salvation with the theory of social change." This mirrors the opposite and equally self-serving conservative ideology that alleges a "permanent nature of things" and "constant limits of human nature," in order to "protect some contingent social structures."[7] Not only, by contrast, do the American prophets express mea culpas in more or less theological senses, they abominate any sacralizing of political systems in the guise of the scientific as they abominate scientific systems in the guise of the political. This, which reflects more than attitude and comes closer to following from their articulation of their life worlds, seems to embody in varying proportions a disdain for systemic pretensions to totalization, an acknowledgment of mystery (including the mystery of iniquity, not a *thing*, yet more and deeper than error), a respect for the unexpectedness of the universe (against the conviction that there is a "permanent nature of things"), and an affection for the universe's less powerful inhabitants.

What then with regard to writing as a career and to the ends of writing and writers' careers? Rousseau had "through accusation and rejection conjugated actively and passively . . . the myth of a social universe [of] fellow men in a perfect transparency of hearts[,] and the myth of an individual experience so rich that it can replace all other possessions."

7. *Ibid.,* 353, 369*n*25.

With the system complete, it is "possible to abandon the roles of . . . the paternal figure [for] life in nature, . . . fusion, . . . the return to the mother."[8] Modern American prophecy *is* demythologizing, as I have tried to show through these three, and specifically includes, albeit in rudimentary form in Mumford, an abjuration of the "transparency" of hearts or language. The very sign is *en procès* between user and addressee, and the *being* of both user and addressee is analogously *en procès,* rather than the sign's constituting the quasi–Bohr atom of Saussure. What the three essayists do invoke, call it coexperience, is much more complex and problematic. That by its nature is no individual experience to replace all other, nor is it quite a possession, as the very dynamism even of Eiseley's image, in "The Last Neanderthal," of key experiences as a limited number of plums to cast for germination or humus makes plain. The three Americans give their voices to writerly silence, to move our readerly silence.

But if Eiseley, the lovingly fathered, can more than the other two suggest natural/maternal *fusion,* the moments are hard to find outside the final autobiographical chapters. There and in the late poems, images of frozen altiplano or of resorption in the eternal storm or of the predatory animal kingdom add turbulence to the movement. Tate, coolly, faintly fathered by a man who knew disgrace, but quasi-paternally brothered, seems in his essays, and in many scores of letters, to be to the end brotherly, if a familial term must be chosen. Admittedly it is tempting to detect a certain mellow fatherliness in his late essay on the southern myth, as his private life exfoliated in his second family even as physical afflictions pressed harder on him. Admittedly, too, his poetry is a separate case. Mumford, virtually fatherless save for his quasi-parental maternal grandfather, seems the compensatorily fatherly legislator even in his autobiographical volumes, which I have scanted: *Sketches from Life, Findings and Keepings, Works and Days.* He winnows and marshals drawings, pictures, manuscripts, and letters, like troops to be drilled and bivouacked. But he does so with some humility and with considerable respect and affection for old loves, for old friends—like "dear Van Wyck"—with whom he had profound disagreements, and for all but the most relentless totalizers. So it obviously is with Tate, in essayistic references and in the often testy but voluminously faithful correspondence—with Kenneth Burke, with Edmund Wilson, with Yvor Win-

8. *Ibid.,* 367, 363–64.

ters,[9] with Karl Shapiro, to name only some of the most notable instances of conflict. So it less obviously is with Eiseley. His gracious way with a great range and number of epistolary petitioners is apparent in the unpublished archival collections at the University of Pennsylvania. He wrote gently of Pierre Teilhard de Chardin as a basically kind and perhaps dotty old man of significant early accomplishment, whereas Mumford came to regard Teilhard as at best an unintentionally dangerous abstractor and totalizer. Eiseley lauded the "rare . . . seer's eyes" of Mumford in an unpublished review of *Technics and Human Development*, a book from which he apparently derived his title *The Invisible Pyramid*. He later quarried bits from the review, on the Palomar eye and microscope eye situating mankind in between, and about a civilization turning "deathward" (see above, pp. 231, 232).

Thereby hangs a tale—but a very complex one, as in the Mandelbrot set or a fractal boundary. Definition through a complex of contrasts with Rousseau can be supplemented by a further definition by contrast with the psychoanalytic complex Paul Ricoeur indicates in the epigraph to Chapter 1. Fevered idolaters and haters of technology alike suffer from distortion or diversion of "enjoyment"—which is to say, love— into power. With that go concordant degradations of language (as concept and usage) to power tool, and all the fears of what is not controlled. Does each of us not love at least the power of taking charge of self, as an adult? But in another text that must have been known and approved by Tate, Eiseley, and even the ambivalently theological Mumford, as well as by Ricoeur, "love casteth out fear."

For the three essayists, nature, in the hazily imaginable sense in which it antedated man, mysteriously eventuated in man and culture. But concurrently man's symbolizing activity—*preeminently for each,*

9. If Tate's letters to Winters and Burke exist, they should certainly be published with their splendid letters to him, which are preserved in the Tate Collection at Princeton University. Lewis Simpson has emphasized a Tate almost desperate because of desire as will, constrained by history. See especially Simpson, "The Critics Who Made Us: Allen Tate," *Sewanee Review*, XCIV (1986), 471–85. Louis D. Rubin, Jr., has, among many other thoughtful points about Tate and his literary associates, put something of that as personal perfectionism. See his *The Wary Fugitives* (Baton Rouge, 1978), and *A Gallery of Southerners* (Baton Rouge, 1982). I grant that aspect of Tate, although it seems less marked in his literary presence than it may have been in his presence to his personal friends. In any case, I am more struck by the prophetic aspect of Tate, with desire as *agape*, constrained by the fallen will to power and its phantasms in history but not hopelessly constrained.

linguistic—made culture and "nature," any nature man can name or know, in ongoing turbulent interaction with the unknown and unnameable. The knowing and naming can only be social and *en procès*.

So would the writers, given life and chance, join in chorus with Rousseau's antagonistic latter-day brother Jean François Lyotard?

> The nineteenth and twentieth centuries have given us as much terror as we can take. We have paid a high enough price for the nostalgia of the whole and the one, for the reconciliation of the concept and the sensible, of the transparent and the communicable experience. Under the general demand for slackening and for appeasement, we can hear the mutterings of the desire for a return of terror, for the realization of the fantasy to seize reality. The answer is: Let us wage a war on totality; let us be witnesses to the unpresentable; let us activate the differences and save the honor of the name.[10]

Yes and no: descant, variation (even of key), rather than plainsong, in this hypothetical chorus. More even than witnesses to the unpresentable, the three Americans present themselves, however paradoxically, as representative of the unpresentable, as continual, never-finished, but demythologized and demythologizing versions of themselves. And they could react with asperity to persons whom they took to be short-circuiting the complex pattern of generating a better self, and thus nature, and thus culture.[11]

Furthermore, Lyotard may be correct that the essay is a postmodern form, whereas the fragment is modern.[12] But the three essayists—Tate less markedly—begin their essays, conduct their essays, end their essays with fuzzy, as if microfragmented, boundaries. Another way to put it is that their boundaries are made nonsimple and nonlinear by the turbulence of presence. The writers do not, for all the occasional combativeness so notable in Tate and Mumford, *end* essays with calls to

---

10. Jean François Lyotard, *The Postmodern Condition: A Report on Knowledge*, trans. Geoff Bennington and Brian Massumi (Minneapolis, 1984), 81–82. Paul Connell drew this to my attention.

11. See the testimony of E. Fred Carlisle in *Loren Eiseley: The Development of a Writer* (Urbana, Ill., 1983), Preface, final chapter; Allen Tate to John L. Stewart, in *L*, Appendix N; Tate's disarming admission in *For.*, Preface; the letters by Eiseley quoted above, Chapter 8; and Eiseley's admissions of fugitive growth, *passim*. For Mumford's somewhat displaced edginess, see *VWB*, letters about the National Academy's recognition of Charles Beard; Thomas S. W. Lewis, "Mumford and the Academy," *Salmagundi*, Summer, 1980 (concerning Modern Language Association scholarly editions).

12. Lyotard, *The Postmodern Condition*, 181.

more or less metaphoric battle. Their tendency in course and in conclusions is to signify care for that which provides their signifiers. That is, they honor nature and historic culture for providing their signifiers, which are both the medium and constraint of their social creativity, something variably, but only partially, distinct from their biological creativity.

Construed from the other side, their signifieds bend with affection toward that which constrains them, conspicuously so in Eiseley's idiosyncratic use of *far*. But through their care, their partly inchoate, desire-powered signifieds may alleviate distorting or reductive abuses of signs in the referential world. Hence, their words and lesser syntactic constructions aggregate with a kind of metatextual coherence in the essays I have construed as dismantlings of mythic phantasms and deathly over-controlling projects.

The witness against tychastic signs entails for all three testimony counter to ignorance of the past or to more insidious notions of the past as tychastic. That in turn goes with the recognition that a tychastic past can yield only a present that is tychastic or entropic. And anyone construing the present as tychastic or entropic will be vulnerable to phantasmal, mythic constructions of a future, demonic or nightmarish. More particularly, concretely, and for notable example, Mumford made an Egypt widely decayed to junior-high-school yawn[13] into a thing of darkness we must acknowledge ours, Tate quarried the cultural past to reinforce and rehabilitate the farmstead and hearth of our corporate language, and Eiseley revivified the very notion and feel of tradition, a notion that has been thought abused by culture or played out, and showed tradition's stunning expansibility with regard to the referential reach of culture and of the very notion and sense of time itself.

Mumford exhibited his own middle-aged Emersonian bias but could be writing for all three in the not-quite-Saussurean terms he employed: "We must throw off much of the burdensome apparatus of our present life: we must break the prevailing images, abandon the glib routines and empty ceremonies: challenge the existing ideological archetypes, and return, as near as possible to the naked person, alone with his cosmic over-self" (*CL*, 252). The iconoclasm of prevailing images includes both stigmatizing certain signs (*efficiency, progress,* asphalt, the UN build-

13. See James Herndon's account of "doing Egypt" in the seventh grade, in *How to Survive in Your Native Land* (New York, 1971).

ing) and breaking signifiers apart from signifieds confusedly attached to them (*communication,* in "The Man of Letters"). By "ideological archetypes" may be understood the phantasmal concepts, notions, and impulses that have been one focus of this study, driven by irrational desires for civic control or Eden without death and sailing under signifiers of convenience.

But "glib routines and empty ceremonies" to be abandoned—that prescription reaches toward the *syntax* of signs and the ontology of reference. Abandoned for what? Empty of what? We must seek to be "alone," Mumford says, with "cosmic over-self," as Emerson too might have said, or Eiseley might have come near to saying, though Tate never. But "*we* must"? He explicitly and each of the others implicitly insists that even the solitary meditation, occupation, or conclusion must be significantly coexperiential. That is what glib routines and ceremonies are empty of—copresence, coexperience. For they reduce the human other to an event in time, or a term in a *linear* equation or process, or an object in space.

Tate might seem to have started far afield, so to speak: "The enemy to all those New Englanders was Nature. . . . The general symbol of Nature, for [Dickinson], is Death, and [Dickinson's] weapon against Death is the entire powerful dumb-show of the puritan theology led by Redemption and Immortality" (*Reac.,* 11–12). One may be struck by the attribution of voiceless silence, and the Otherness of this "Nature." Yet one notices the dynamism implicit in *weapon* and *powerful* and *led by:* the suggestion of resourceful Emily Dickinson executing endless and wonderfully various, never quite congruent figure eights around the paired notions of death and God.

The twentieth-century three loop their figures around masked surrogates of death, unmasked by reference to garden, animal, sheeted-bedside, or hearth-, table-, square-, cemetery-, or textside, scenes of copresence. Tate wrote, much later, "We fervently hope that at this moment of decline in civilized intercourse, we can preserve together, by creating anew, that order of intelligence without which mankind will not need to go to the moon. If we don't have that order of intelligence, we shall have a sufficient desert right here" (*Mem.,* 207).

Infinitely complex is the turbulence of creation occurring amid decline that is identified as deathly. Mumford cites Walt Whitman's linearly disjunctive remark that we "convince not by our argument but by our presence." But the essays of Mumford, Tate, and Eiseley are profoundly a literature of presence, in a Heideggerian sense of being which endures us-ward.[14] Obviously it is necessary to be careful with these three writers—as with most twentieth-century American prophets, though not with Europeans such as Jacques Ellul—about equating presence with the presence of God or of the god, which might have the effect of putting divinity out of play except in the sense of that in which one places one's ultimate trust. It would be more apposite to think of a god who may appear in the "least of these," as, for instance, in Eiseley's hobo, and fox cub. Like any careful deconstructor, one must avoid the linearity of making an exhaustive alternative of "full presence or utter absence." That would be bad logic, implicated in the fallacy of the slippery slope. More existentially, it might bespeak a tragic sense of life, the distance of which from the three Americans' writings can be judged by the degree of contrast with Donne's conclusion to his "Lecture upon the Shadow":

> Love is a dawning, or a full constant light,
> And his first minute after noon is night.

Love in a multitude of registers, love diversely and turbulently implicated in darkened linguistic and perceptual structures of a technocratic world, has animated the three writers' cautionary messages from night places of the world, the study, and the heart. Love is in this world never perfect, usually difficult, and not always even possible, to human frailty. But as Tate's remarks about the hope of preserving and renewing mankind's intelligence can fairly suggest, hope underlies the writing and promulgation of the essays of all three. And their hope is more than merely the wish to reach across formidable distance; it is the wish somehow to overpass formidable obstacles, to brotherly or sisterly coexperiences:

> —a world where techniques are paramount
> is a world given over to desire and fear;
> because every technique is there to serve
> some desire or some fear.
> It is perhaps characteristic of Hope to be unable
> either to make direct use of any technique
> or to call it to her aid.

14. Martin Heidegger, *The Question Concerning Technology, and Other Essays,* trans. William Lovitt (New York, 1977), 9.

Hope is proper to the unarmed;
it is the weapon of the unarmed;
or (more exactly) it is the very opposite of a weapon
and in that, mysteriously enough, its power lies.[15]

In the nonexecutive essays that this volume has sounded, hope itself is the action and model of Lewis Mumford, Allen Tate, and Loren Eiseley, watchers and heralds in the night.

15. Gabriel Marcel, *Homo Viator,* trans. Emma Cranfurd (New York, 1962), March 17, 1931.

SELECTED BIBLIOGRAPHY

Agar, Herbert, and Allen Tate, eds. *Who Owns America? A New Declaration of Independence.* Boston, 1936.

Angyal, Andrew. *Loren Eiseley.* Boston, 1983.

Bachelard, Gaston. *The Poetics of Space.* Translated by Maria Jolas. Boston, 1964.

Barthes, Roland. *Mythologies.* Selected and translated by Annette Lavers. New York, 1972.

Bateson, Gregory. *Steps to an Ecology of Mind.* New York, 1972.

Benjamin, Walter. *Illuminations.* Translated by Harry Zohn. New York, 1969.

Berle, Adolf A., Jr., and Gardner Means. *The Modern Corporation and Private Property.* New York, 1933.

Berne, Eric. *Games People Play.* New York, 1964.

Bettelheim, Bruno. *Symbolic Wounds: Puberty Rites and the Envious Male.* Glencoe, Ill., 1954.

Blackmur, R. P. "San Giovanni in Venere: Allen Tate as Man of Letters." In *Allen Tate and His Work,* edited by Radcliffe Squires. Minneapolis, 1972.

Bové, Paul. "Agriculture and Academe: America's Southern Question." *Boundary 2,* XVI (1988), 169–95.

Brooks, Van Wyck. *An Autobiography.* New York, 1965.

Buffington, Robert. "Allen Tate: Society, Vocation, Communion." *Southern Review,* XVIII (1982), 62–72.

Bunn, James. *The Dimensionality of Signs, Tools, and Models.* Bloomington, Ind., 1981.

Carlisle, E. Fred. *Loren Eiseley: The Development of a Writer.* Urbana, Ill., 1983.

Cassirer, Ernst. *Language.* New Haven, 1953. Vol. I of Cassirer, *The Philosophy of Symbolic Forms.* 3 vols.

Conrad, David R. *Education for Transformation: Implications in Lewis Mumford's Ecohumanism.* Palm Springs, Calif., 1976.

Core, George. "Agrarianism, Criticism, and the Academy." In *A Band of Prophets,* edited by William C. Havard and Walter Sullivan. Baton Rouge, 1982.

Cowan, Louise. *The Fugitive Group: A Literary History.* Baton Rouge, 1959.

Cox, Harvey. *The Feast of Fools: A Theological Essay on Festivity and Fantasy.* Cambridge, Mass., 1969.

Dennison, George. *The Lives of Children: The Story of First Street School.* New York, 1969.

Derrida, Jacques. "Living On. Border Lines." In *Deconstruction and Criticism,* edited by Harold Bloom *et al.* New York, 1979.

———. *Margins of Philosophy.* Translated by Alan Bass. Chicago, 1982.

———. *Of Grammatology.* Translated by Gayatri Spivak. Baltimore, 1976.

———. *Speech and Phenomena, and Other Essays on Husserl's Theory of Signs.* Translated by David B. Allison. Evanston, Ill., 1973.

———. *Writing and Difference.* Translated by Alan Bass. Chicago, 1978.

Donoghue, Denis. "On Allen Tate." In *Allen Tate and His Work: Critical Evaluations,* edited by Radcliffe Squires. Minneapolis, 1972.

Douglass, Paul. *Bergson, Eliot, and American Literature.* Lexington, Ky., 1986.

Drucker, Peter. *The Age of Discontinuity: Guidelines to Our Changing Society.* New York, 1969.

Dupree, Robert S. *Allen Tate and the Augustinian Imagination.* Baton Rouge, 1983.

Eiseley, Loren. *All the Night Wings.* New York, 1979.

———. *All the Strange Hours: An Excavation of a Life.* New York, 1975.

———. *Another Kind of Autumn.* New York, 1977.

———. *Darwin and the Mysterious Mr. X: New Light on the Evolutionists.* New York, 1979.

———. *Darwin's Century: Evolution and the Men Who Discovered It.* New York, 1958.

———. "The Ethic of the Group." In *Social Control in a Free Society,* edited by Robert E. Spiller. Philadelphia, 1960.

———. *The Firmament of Time.* New York, 1960.

———. Foreword to *Not Man Apart: Lines from Robinson Jeffers, with Photographs of the Big Sur Coast,* ed. David Brower (San Francisco, 1965).

———. "Fossil Man and Human Evolution." In *Yearbook of Anthropology,* edited by William L. Thomas, Jr. New York, 1955.

———. *The Immense Journey.* New York, 1957.

———. *The Innocent Assassins.* New York, 1973.

———. Introduction to *The Shape of Likelihood: Relevance and University,* edited by Taylor Littleton. University, Ala., 1971.

———. Introduction to *The Voyage to Arcturus,* by David Lindsay. New York, 1963. Reprint of 1st ed., London, 1920.

————. *The Invisible Pyramid*. New York, 1979.

————. *The Lost Notebooks of Loren Eiseley*. Edited by Kenneth Heuer. Boston, 1987.

————. *The Man Who Saw Through Time*. New York, 1973.

————. *The Night Country*. New York, 1971.

————. *Notes of an Alchemist*. New York, 1972.

————. *The Star Thrower*. New York, 1978.

————. *The Unexpected Universe*. New York, 1969.

Eiseley, Loren, and Eliot Porter. *Galápagos: The Flow of Wilderness*. London, 1968.

Eisenstein, Elizabeth. *The Printing Press as an Agent of Change in Early Modern Europe*. 2 vols. New York, 1979.

Elder, Frederick. *Crisis in Eden: A Religious Study of Man and Environment*. Nashville, 1970.

Ellul, Jacques. *The Technological System*. Translated by Joachim Neugroschel. New York, 1980.

Erikson, Erik. *Childhood and Society*. New York, 1950.

————. *Identity, Youth, and Crisis*. New York, 1968.

Fain, John Tyree, and Thomas Daniel Young. *The Literary Correspondence of Donald Davidson and Allen Tate*. Athens, Ga., 1974.

Falck, Colin. *Myth, Truth, and Literature: Towards a True Post-Modernism*. Cambridge, Eng., 1989.

————. "Saussurian Theory and the Abolition of Reality." *Monist*, LXIX (1986), 133–45.

Fallwell, Marshall, *et al. Allen Tate: A Bibliography*. New York, 1969.

Fisch, Harold. *A Remembered Future: A Study in Literary Mythology*. Bloomington, Ind., 1984.

Fleming, Rudd. "Dramatic Involution: Tate, Husserl, and Joyce." *Sewanee Review*, LX (1952), 445–64.

Ford, Ford Madox. *Letters*. Edited by Richard M. Ludwig. Princeton, 1965.

Foucault, Michel. *The Archeology of Knowledge*. Translated by A. M. Sheridan. New York, 1972.

————. *Language, Counter-Memory, Practice: Selected Essays and Interviews*. Translated and edited by Donald Bouchard and Sherry Simon. Ithaca, N.Y., 1977.

————. *The Order of Things: An Archaeology of the Human Sciences*. New York, 1970.

Gadamer, Hans-Georg. "The Eminent Text and Its Truth," "Responses," "Final Observations." In *The Horizon of Literature*, edited by Paul Hernadi. Lincoln, Neb., 1982.

Gerber, Leslie E., and Margaret McFadden. *Loren Eiseley*. New York, 1983.

Girard, René. *Violence and the Sacred.* Translated by Patrick Gregory. Baltimore, 1977.

———. *To Double Business Bound: Essays on Literature, Mimesis, and Anthropology.* Baltimore, 1978.

Gleick, James. *Chaos: Making a New Science.* New York, 1987.

Handy, William J. *Kant and the Southern New Critics.* Austin, Tex., 1963.

Heidegger, Martin. *Being and Time.* Translated by John Macquarrie and Edward Robinson. New York, 1962.

———. *The Question Concerning Technology, and Other Essays.* Translated by William Lovitt. New York, 1977.

Heidtmann, Peter. "An Artist of Autumn." *Prairie Schooner,* LXI (1987), 45–56.

Herndon, James. *How to Survive in Your Native Land.* New York, 1971.

Holloway, Ralph L. "The Poor Brain of *Homo sapiens neanderthalensis:* See What You Please." In *Ancestors: The Hard Evidence,* edited by Eric Delson. New York, 1985.

Hughes, Michael, ed. *The Letters of Lewis Mumford and Frederic J. Osborn: A Transatlantic Dialogue, 1938–1970.* New York, 1972.

Ihde, Don. *Existential Technics.* Albany, N.Y., 1983.

I'll Take My Stand. By Twelve Southerners. 1931; rpr. with introduction by Louis D. Rubin, Jr., Baton Rouge, 1977.

Kerrigan, William. *The Prophetic Milton.* Charlottesville, Va., 1974.

Knapp, Bettina Liebowitz, ed. *The Lewis Mumford/David Liebovitz Letters.* Troy, N.Y., 1983.

Kristeva, Julia. "The Pain of Sorrow in the Modern World: The Works of Marguerite Duras." *PMLA,* CII (1987), 138–51.

Krois, John Michael. "Ernst Cassirer's Theory of Technology and Its Import for Social Philosophy." *Research in Philosophy and Technology,* V (1982), 209–22.

Kronick, Joseph. *American Poetics of History: From Emerson to the Moderns.* Baton Rouge, 1984.

Kubler, George. *The Shape of Time: Remarks on the History of Things.* New Haven, 1962.

Laing, R. D. *The Divided Self.* London, 1960.

Lifton, Robert Jay. *The Broken Connection.* New York, 1979.

McKnight, Stephen A. *Sacralizing the Secular: The Renaissance Origins of Modernity.* Baton Rouge, 1989.

Makowsky, Veronica. *Caroline Gordon: A Biography.* New York, 1989.

Marcel, Gabriel. *Being and Having.* Translated by Katherine Farrer. 1949; New York, 1965.

———. *Creative Fidelity.* Translated by Robert Rosthal. New York, 1964.

Marin, Louis. "Writing History with the Sun King: The Traps of Narrative." In *On Signs,* edited by Marshall Blonsky. Baltimore, 1985.

Marshack, Alexander. *The Roots of Civilization: The Cognitive Beginnings of Man's First Art, Symbol, and Notation.* New York, 1972.

Martz, Louis. *The Paradise Within.* New Haven, 1964.

———. *The Poem of the Mind: Essays on Poetry, English and American.* New York, 1966.

———. *The Poetry of Meditation.* New Haven, 1954.

Meiners, R. K. *The Last Alternative: A Study of the Works of Allen Tate.* Denver, 1963.

Molella, Arthur P. "Inventing the History of Invention." *Invention and Technology,* Spring–Summer, 1988, pp. 23–30.

Morris, Wright. *A Cloak of Light: Writing My Life.* New York, 1985.

Mumford, Lewis. *The City in History.* New York, 1961.

———. *The Condition of Man.* New York, 1944. Vol. III of Mumford, *The Renewal of Life.* 4 vols.

———. *The Conduct of Life.* New York, 1951. Vol. IV of Mumford, *The Renewal of Life.*

———. *The Culture of Cities.* New York, 1938. Vol. II of Mumford, *The Renewal of Life.*

———. *Findings and Keepings.* New York, 1975.

———. *The Highway and the City.* New York, 1963.

———. *My Works and Days: A Personal Chronicle.* New York, 1979.

———. *The Pentagon of Power.* New York, 1970. Vol. II of Mumford, *The Myth of the Machine.* 2 vols.

———. *Sketches from Life: The Early Years.* New York, 1982.

———. *The South in Architecture.* New York, 1941.

———. *The Story of Utopias.* New York, 1922.

———. *Technics and Civilization.* New York, 1934. Vol. I of Mumford, *The Renewal of Life.*

———. *Technics and Human Development.* New York, 1967. Vol. I of Mumford, *The Myth of the Machine.*

———. "Technics and the Nature of Man." In *Philosophy and Technology,* edited by Carl Mitcham and Robert Mackey. New York, 1983.

Newman, Elmer S. *Lewis Mumford: A Bibliography, 1914–1970.* New York, 1971.

Olson, Charles. *The Maximus Poems.* Berkeley, Calif., 1983.

Ong, Walter. *In the Human Grain.* New York, 1967.

———. *Interfaces of the Word.* Ithaca, N.Y., 1977.

———. *Orality and Literacy: The Technologizing of the Word.* New York, 1982.

———. "Samuel Johnson and the Printed Word." *Review*, X (1988), 97–112.

Ortega y Gasset, José. *The Dehumanization of Art, and Other Writings on Art and Culture*. Translated by Willard Trask. Princeton, 1948.

———. *History as a System, and Other Essays Toward a Philosophy of History*. Translated by Helen Weyl *et al.* New York, 1941.

———. *Man and People*. Translated by Willard Trask. New York, 1957.

Pannenberg, Wolfhart. "The Question of God." *Interpretation*, XXI (1967), 289–314.

Peirce, Charles Sanders. *Scientific Metaphysics*, esp. pars. 47–65. Cambridge, Mass., 1935. Vol. VI of Peirce, *Collected Papers*. 8 vols.

Piaget, Jean. *Structuralism*. Translated and edited by Chaninah Maschler. New York, 1970.

Poggi, Gianfranco. *Calvinism and the Capitalist Spirit: Max Weber's "Protestant Ethic."* Amherst, Mass., 1983.

Polanyi, Michael. *The Tacit Dimension*. Garden City, N.Y., 1966.

Quinones, Ricardo J. *The Renaissance Discovery of Time*. Cambridge, Mass., 1972.

Ricoeur, Paul. *The Conflict of Interpretations: Essays in Hermeneutics*. Evanston, Ill., 1974.

———. "Metaphor and the Main Problem of Hermeneutics." *New Literary History*, VI (1974), 95–110.

Riddel, Joseph. "Decentering the Image: The 'Project' of American Poetics?" In *Textual Strategies: Perspectives in Post-Structuralist Criticism*, edited by Josué Harari. Ithaca, N.Y., 1979.

Ross, Stephen. "Oratory and the Dialogical in *Absalom, Absalom!*" In *Intertextuality in Faulkner*, edited by Michel Gresset and Noel Polk. Jackson, Miss., 1985.

Rubin, Louis D., Jr. *A Gallery of Southerners*. Baton Rouge, 1982.

Said, Edward. *Beginnings: Intention and Method*. Baltimore, 1975.

———. *Joseph Conrad and the Fictions of Autobiography*. Cambridge, Mass., 1966.

———. "Reflections on Recent American 'Left' Literary Criticism." In *The Question of Textuality: Strategies of Reading in Contemporary American Criticism*, edited by William V. Spanos, Daniel O'Hara, and Paul Bové. Bloomington, Ind., 1982.

Schell, Jonathan. *The Time of Illusion*. New York, 1976.

Schneidau, Herbert N. "The Bible Under Attack." *Religion and Intellectual Life*, VI (1989), 193–203.

———. *Sacred Discontent: The Bible and Western Tradition*. Baton Rouge, 1976.

See, Fred G. *Desire and the Sign in American Literature*. Baton Rouge, 1987.

———. "'Writing So as Not to Die': Edgar Rice Burroughs and the West Beyond the West." *Melus,* II (1984), 59–72.

Serres, Michel. *Hermes: Literature, Science, Philosophy.* Edited by Josué Harari and David F. Bell. Baltimore, 1982.

Simpson, Lewis. *The Brazen Face of History: Studies in the Literary Consciousness in America.* Baton Rouge, 1980.

———. "The Critics Who Made Us: Allen Tate." *Sewanee Review,* XCIV (1986), 471–85.

———. *The Dispossessed Garden: Pastoral and History in Southern Literature.* Athens, Ga., 1975.

———. "Eric Voegelin and the Story of the 'Clerks.'" *Sewanee Review,* XCVII (1989), 2–29.

———. Foreword to *The Literary Correspondence of Donald Davidson and Allen Tate,* edited by John Tyree Fain and Thomas Daniel Young. Athens, Ga., 1974.

———. Introduction to *Still Rebels, Still Yankees, and Other Essays,* by Donald Davidson. Baton Rouge, 1972.

———. *The Man of Letters in New England and the South: Essays on the History of the Literary Vocation in America.* Baton Rouge, 1973.

———. "The Southern Republic of Letters and *I'll Take My Stand.*" In *A Band of Prophets,* edited by William C. Havard and Walter Sullivan. Baton Rouge, 1982.

Singal, Daniel. *The War Within: From Victorian to Modernist Thought in the South, 1919–1945.* Chapel Hill, N.C., 1982.

Slater, Philip. *The Pursuit of Loneliness: American Culture at the Breaking Point.* Rev. ed. Boston, 1976.

Spiller, Robert E., ed. *The Van Wyck Brooks–Lewis Mumford Letters: The Record of a Literary Friendship, 1921–1963.* New York, 1970.

Squires, Radcliffe. *Allen Tate: A Literary Biography.* New York, 1971.

———, ed. *Allen Tate and His Work: Critical Evaluations.* Minneapolis, 1972.

Starobinski, Jean. "The Accuser and the Accused." *Daedalus,* CXVII (1988), 345–70. Originally in *Daedalus,* CVII (1978).

———. *The Invention of Liberty.* New York, 1979.

Stewart, John L. *The Burden of Time: The Fugitives and Agrarians.* Princeton, 1965.

Sullivan, Walter. *Allen Tate: A Recollection.* Baton Rouge, 1989.

Szondi, Peter. *On Textual Understanding, and Other Essays.* Translated by Harvey Mendelsohn. Minneapolis, 1986.

Tate, Allen. *Collected Essays.* Denver, 1959.

———. *The Forlorn Demon: Didactic and Critical Essays.* Chicago, 1953.

———. *The Hovering Fly, and Other Essays.* Cummington, Mass., 1948.

———. *Jefferson Davis—His Rise and Fall: A Biographical Narrative.* New York, 1929.

————. "Kenneth Burke and the Historical Environment." *Southern Review,* II (1936), 363–72.

————. *Memoirs and Opinions, 1926–1974.* Chicago, 1975.

————. *The Poetry Reviews of Allen Tate, 1924–1944.* Edited by Ashley Brown and Frances Neel Cheney. Baton Rouge, 1983.

————. *Reactionary Essays on Poetry and Ideas.* New York, 1936.

————. *Reason in Madness.* New York, 1941.

————. *Stonewall Jackson: The Good Soldier.* New York, 1928.

————. "The Unliteral Imagination; or, I, Too, Dislike It." *Southern Review,* n.s., I (1965), 530–42.

Tillich, Paul. *The Courage to Be.* New Haven, 1952.

Toulmin, Stephen, and June Goodfield. *The Discovery of Time.* New York, 1965.

Voegelin, Eric. *Anamnesis.* Notre Dame, Ind., 1978.

————. *The New Science of Politics: An Introduction.* Chicago, 1952.

Waldron, Ann. *Close Connections: Caroline Gordon and the Southern Renaissance.* New York, 1988.

Young, Thomas Daniel, and Elizabeth Sarcone, eds. *The Lytle-Tate Letters: The Correspondence of Andrew Lytle and Allen Tate.* Jackson, Miss., 1987.

INDEX

Abraham and Isaac, 207
Abstraction: within technology and
  capitalism, 8, 9; as myth, 17, 19; and
  power, 24, 29, 80; as nihilistic, 37;
  celebrated by Mumford, 43, 84;
  ambivalence of, 82, 87; as taxonomic,
  89; in exploration, 90; as evidence of
  myth of the machine, 91; and motion,
  93; as "death," 135; bodily opposition
  to, 141; as "magic," 165; as fury,
  178; tychastic or taxonomic, 185;
  mentioned, 42, 68, 91, 95, 109, 116,
  118, 123, 138, 154
Action as responsible decision, 215–16
Adams, Henry, 65, 97, 100, 129
Adorno, Theodor, 8
Africa, 125
Agar, Herbert, 121
Agrarianism: Tate on, 132; as resisting
  closure, 133; Bové's attack on, 135; of
  Tate and Gordon, compared, 136n8;
  Tate's 1933 definition of, 147; as
  means, 149; mentioned, 133
Alexander of Macedon, 211
Allegory, 22, 42, 44, 62–65, 162,
  186n26, 218
Amenia, New York, 15, 46, 86, 106
Americanism: and life, 3, 77, 131, 134;
  Mumfordian, 33, 55, 60, 62, 72–73,
  96, 106; and man in between,
  107–108; Tatian, 120, 136–37,

142–43, 146, 154–55; Eiseleyan, 188,
  200–201, 208, 213, 228; as
  coexperience, 276
American Review, 110n2, 120
American Scholar, 215, 253
American Scientist, 215
American Telephone and Telegraph
  Company, 132
Amos, 64, 72, 78, 194
Amsterdam, 24
Analogies: competing, 168–69;
  mechanical, 171
Angyal, Andrew, 195n7, 211n28,
  221n39, 222n42, 232n6, 234n9,
  238n15, 250n6
Apollo, 77
Aquinas, Saint Thomas, 17, 50, 124
Arendt, Hannah, 66
Argument from design, 244
Aristotle, 34, 176
Arnold, Matthew, 123, 128, 175
Atavism, 113–14
Atum-Re, 92
Auden, W. H., 152, 164, 165, 203, 220
Augustine, Saint, 2, 37, 76, 99, 100, 149,
  156, 158, 160, 164, 172, 189, 218,
  221, 223, 273
Authority: as center, 131; of human
  subject, 225
Automation, 103
Avalon, 237. *See also* Eden

291